The Institute of British Geographers
Special Publications Series

19   Technical Change and Industrial Policy

The Institute of British Geographers
Special Publications Series

Technical Change
and Industrial Policy

Edited by
Keith Chapman and Graham Humphrys

Basil Blackwell

British Library Cataloguing in Publication Data

Technical change and industrial policy.—
   (Institute of British Geographers special
   publication, ISSN 0073–9006; v. 19)
   1. Industry   2. Technological innovations
   3. Geography, Economic
   I. Chapman, Keith   II. Humphrys, Graham
   III. Series
   338'.06   HC79.T4

ISBN 0-631-15215-6

Library of Congress Cataloging in Publication Data

Technical change and industrial policy.

   (The Institute of British Geographers special
publications series; 19)
   Selected papers presented at joint meetings of the
Institute of British Geographers and the Canadian
Association of Geographers, held at Swansea, 1985.
   Bibliography: p.
   Includes index.
   1. Technological innovations. I. Chapman, Keith.
II. Humphrys, Graham. III. Series: Special publication
(Institute of British Geographers); no. 19.
T173.8.T34 1987   338.9'26 87–1856

ISBN 0-631-15215-6

Typeset by System 4 Associates, Gerrards Cross, Buckinghamshire
Printed in Great Britain by Page Brothers (Norwich) Ltd

Contents

vi *Contents*

Part III Technical Change and Spatial Policy

Preface

Industrial geography has been one of the most dynamic areas of the discipline in recent years. Its strength is reflected in the impressive output of published research and its importance is emphasized by the relevance of these contributions to the policy debate at a time of rapid industrial change. The increasing scale of industrial organization has encouraged the adoption of wider perspectives which explicitly acknowledge the inter-relationships between, for example, industrial growth in one place and decline in another. Increasingly, these inter-relationships extend across national frontiers. In these circumstances, it is important that academic research becomes less parochial in its concerns and any forum which encourages the exchange of ideas between practitioners in different countries must be regarded as desirable. Thus the Industrial Activity and Area Development study group of the Institute of British Geographers and its counterpart within the Canadian Association of Geographers have formally met at two year intervals, starting in Edinburgh in 1981, moving to Calgary/Edmonton in 1983 and Swansea in 1985. This volume is based on selected contributions at the Swansea meeting and is the third in the series (Collins, 1982; Barr and Waters, 1984). It actually includes less than half the papers presented in Swansea where the chosen theme, 'Technical change in industry – spatial policy and research implications', proved successful in attracting participants. The contributions not included in this volume were important elements of the programme and are gratefully acknowledged. Thanks are also due to those who are included for responding promptly and with good grace to editorial suggestions. Finally, Mike Clark and David Pinder played an important part in getting this book off the ground as a result of their negotiation, on behalf of the Institute of British Geographers, with the publishers.

Keith Chapman
Graham Humphrys

REFERENCES

Barr, B. M. and Waters, N. M. (1984) *Regional diversification and structural change*, Tantalus Research Ltd., Vancouver BC.
Collins, L. (1982) *Industrial decline and regeneration*, Department of Geography and Centre of Canadian Studies, Edinburgh.

1

Introduction

K. Chapman and G. Humphrys

The period since 1970 has witnessed major advances in industrial geography. Despite wide differences of approach, method and philosophical perspective, there has been a general recognition that neo-classical location theory, which represented the traditional point of departure in the analysis of industrial location, is far too narrow a foundation for contemporary research. The neo-classical tradition encouraged geographers to seek explanations for the distribution of manufacturing in the environment external to industry, notably in spatial variations in the cost and availability of the various factors of production as defined in classical economics. Dissatisfaction arose from the lack of practical applicability of both the theories and the findings of this work and its inadequacy as a source of explanation in so many cases. As a result the research emphasis shifted. It became focused on extending the traditional approach by including a far wider range of external influences. These included the idea that industrial distributions conform to the structure of space as interpreted by core-periphery theory, the importance of the temporal evolution of regional structure, the influence of the product cycle and the significance of technical change in explaining spatial aspects of industrial change. A further development was the use of the structuralist or radical approach. In this, the reciprocal relationship between industrial location and the structure of society is emphasized. Another line of research explored the reasons why the same external environment may produce different spatial patterns between firms within the same industry. This stressed the significance of such features as type of ownership, form of business organization and corporate spatial structure as influences upon the distribution of manufacturing. This attempt to view the industrial location problem within an organizational context was itself part of a move towards a behaviourist approach which occurred in other branches of human geography.

The paradigm shift was largely complete by the early 1980s. By then the results of research work being published looked very different from those which had been appearing twenty years earlier. The key to the fundamental difference between the two periods lies in the differences in the explanations provided for spatial patterns and spatial change. The nature of the information used to answer the initial, basically descriptive, geographical questions 'where is it and what is it like?' provide good examples. Where the explanation of the patterns is based in the internal environment, the descriptions require data about the geography of the company concerned. In the case of the multiplant enterprise, these requirements include such things as the nature and age of its capital equipment and the role and relationships of its various facilities. Earlier studies would have emphasized the variations in the external environment, such as labour costs or infrastructure availability. The different explanations also call for different policies to achieve specified objectives. There is less confidence that changes in the external environment, epitomised by infrastructure investment in transport facilities, will stimulate industrial development in the desired locations. Furthermore, the insights into the internal structure of firms and industries provided by recent research efforts have emphasized that cetain types of investment are more desirable than others as far as their potential impacts upon the economic development of communities and regions are concerned.

Although the basic objectives of industrial geography, to describe and explain the spatial distribution of manufacturing and related economic activities, remains the same in the 1980s as it was in the 1960s, the specific concerns are very different. During the 'boom' of the 1960s, it seemed appropriate to focus upon, for example, industrial movement and the decision-making processes involved in the choice of location for new plant. In the 1970s and 1980s, processes of decline have been the dominant influence upon the distribution of manufacturing within most developed economies, and this has been reflected in various attempts to understand the spatial dimensions of these processes (Massey and Meegan, 1982; Townsend, 1982). The role of technology is a unifying theme relevant to an understanding of both industrial growth and industrial decline. Thus obsolescence and innovation represent opposite ends of a spectrum epitomised, at the industry scale, by such epithets as 'sunset' and 'sunrise'. A common objective of policy makers at various levels of government is to encourage and to attract those activities which, by virtue of their technological characteristics, are perceived to offer the prospect of future growth in output and, especially, employment. Such an objective often rests upon a kind of technological determinism fostered by life-cycle models applied to products, industries and regions. It also betrays a belief in the benefits of 'high technology' which is not always justified. This volume explores these and other issues regarding the role of 'technology' in the interpretation of spatial patterns of industrial growth and decline. It is divided into three sections. The first is concerned with various aspects of the nature of technical change as a phenomenon. The second evaluates its role in the restructuring of 'traditional' industries whilst the third reviews some of its implications for industrial and regional development policies.

NATURE AND SPATIAL DIMENSIONS OF TECHNICAL CHANGE

The four chapters which comprise the first part of the volume deal with technical change at a variety of scales. Despite the sharp contrast between their macro- and micro-levels of analysis, the first two, by Freeman and McArthur respectively, are complementary in the sense that they are essentially concerned with the diffusion of innovations. These chapters provide a useful context for the contributions of Wood and Harrison which exemplify some of the impacts of technical change upon the employment structure of national economies and upon the occupational structure of an individual industry.

In chapter two Freeman provides a review of the role of technical change in the explanation of long waves at the national and international level. Recent thinking on this topic up to 1985 is included, with the significance of the diffusion process in accounting for cyclic phenomena given prominence. Of particular interest is the reference to the rediscovery of Marx's comments on the influence of changes in the rate of profit on replacement and new investment from volumes 2 and 3 of *Das Kapital*. This Freeman ranks with the rediscovery of Keynes's comments (1930) on the stimulus of new technology in relation to waves of investment and his acceptance of Schumpeter's explanations of business cycles. The analysis by Freeman of technological innovations leads him to propose the following threefold taxonomy: 1 incremental innovations, 2 radical innovations and 3 technological revolutions. The latter Freeman suggests, correspond to 'creative gales of destruction' at the heart of Schumpeter's long-wage theory. All three need to be tested to assess their value in the study of geographical change. Freeman goes on to explore the implications of the information technology paradigm for the fifth Kondratiev wave. In this he emphasizes the need for institutional change in regional development policy and, in particular, the need to replace measures used in the 1950s and 1960s, which are now inappropriate. In this he looks at the way in which the Japanese are developing their approach to regional development. For those geographers coming to this field for the first time, chapter 2 is thus bristling with ideas for further work.

The work of McArthur presented in chapter 3, moves to the micro scale.

This study uses the introduction of the colour scanner as an innovative technology in the UK printing industry as the example. Those who accept uncritically the idea of the broad sweeps of cycles based upon innovation will find their enthusiasm curbed perhaps by the complexity of the interactions between producers and users of new technology which chapter 3 explores. The implications of the findings for prediction and for regional development raise a range of questions which, as the author points out, are not always easy to answer.

Chapters 4 and 5 by Wood and Harrison complement each other. Chapter 4 deals with structural changes accompanying the information revolution, while chapter 5 is concerned with the effects of new technology on one industry in a local area. Wood follows the thesis presented by Freeman in chapter 2, that the information revolution provides the basis for a fifth Kondratiev cycle of major change. To support his thesis he focuses upon the role of producer services in the process of economic growth in Canada. The study demonstrates the need for a change in thinking about the significance of services within the advanced economies, which is now emerging as a major research theme amongst economic geographers. Certainly he shows that technological change is having an enormous impact in this sector and through this, upon the structure of the Canadian economy as a whole.

Harrison, by working on the impact of new technology in the shipbuilding industry in Belfast, provides the detailed empirical analysis which is a necessary basis for broader generalizations. His special interest is in the changes in occupational structure as the technology of the industry changed in the 1960s and 1970s. Using his results, in his concluding section he examines the variations in occupational structure in shipbuilding and marine engineering that can be found in other regions of the UK. He also demonstrates the way in which further studies of technology/occupation change relationships in specific sectors can be integrated with wider debates about the role of job change and the redivision of labour in the processes of class and industrial restructuring in advanced capitalist societies.

TECHNICAL CHANGE AND INDUSTRIAL RESTRUCTURING

The three chapters in this second part of the book are all concerned with different aspects of restructuring resulting from technical change. Restructuring is at the heart of many of the changes in the spatial distribution of industry and of the changing fortunes of particular areas and so is of considerable practical interest and concern. The case studies are important because they are all essentially continental in scale and, in their concern with oil refining, iron and steel and automobile manufacture, they emphasize that 'maturity' in industrial terms is not necessarily synonymous with technological obsolescence.

Pinder and Hussain begin chapter 6 by recognizing four alternative strategies that different firms in a mature industry can adopt in response to technological change. These are offensive strategies, adoptive strategies, imitative strategies and nonresponsive strategies. Implicit in such a classification is the importance of the internal environment and the need to disaggregate the industry, often to the individual firm level, if valid explanations are to be achieved. They go on to point out the importance of distinguishing product innovation from process innovation; the need to distinguish strategies for technical change at the level of the firm from those at the level of the plant; the value of having a theoretical perspective on long-term technological planning by enterprises; and lastly, the need to distinguish short-term from long-term responses. They then present their analysis of the changes in the technological profile of the oil refining industry in Western Europe between 1979 and 1985, paying particular attention to the spatial dimensions of this profile as it may affect any further reductions in capacity.

Holmes focuses upon the Canadian automobile industry. He sees rapid technical change in both product and process as the main reason for the dramatic changes which have occurred in the structure and organization of this industry. He then points out the significant implications which this conclusion has for the future of the automobile industry at both the world and North American levels. The contention is that a new model of production is emerging, which contrasts with and replaces the Fordist model which has operated in the past. The new model involves substantial changes in production technology, in the shop-floor organization of the labour force and in the broader organization of production systems. Because it is widely applicable to other sectors of manufacturing, he suggests that this is likely to have a profound influence upon the location of all industry and on location theory.

Bradbury's analyis of technological change in the North American steel industry explicitly distinguishes between the impact of external conditions upon and the nature of internal responses within, production systems. Technical change is regarded as only one of a number of variables having complex cause and effect relationships with industrial restructuring. The arguments are illustrated with reference to recent changes in the steel industry, first at an international scale and then with reference to the US and to Canada. The changes are regarded as typical of late capitalism and the approach is set firmly within the structuralist tradition.

TECHNICAL CHANGE AND SPATIAL POLICY

The four chapters in the final section are especially concerned with the implications of technical change in industry for spatial policy. This is an area of concern to governments at various levels as a result of the failure of existing policies to cope with the problems posed by the newly emerging industrial geography of the end of the twentieth century. These contributions add to the growing literature in this field in a number of ways.

In chapter 9, Charles looks at the implications of external control, and in particular at the degree of technological autonomy, as one dimension of the branch plant syndrome in the UK electonics industry. Case studies of a number of electronics companies operating in Britain are used to demonstrate the opportunities for the geographical dispersal of both research and development (R&D) and manufacturing activities presented by the emergence of decentralized corporate structures. However, the exploitation of these opportunities may be constrained by labour supply considerations which seem to be critical in the initial development of the electronics industry. Nevertheless, Charles suggests that there is a need for government to demonstrate the possibilities for dispersion, by example, using those parts of the industry which are directly under its control.

The relationship of technology, dependence and regional development is further explored by Todd and Simpson in Chapter 10 with reference to the aerospace industry. The similarity of experiences in Manitoba and Northern Ireland emphasizes the high risks which are involved for public money when it is associated with strategies for development based upon specific 'high technology' sectors. In particular these experiences suggest that picking winners is not easy. If it is to be successful it requires co-ordination of regional and national policies and the integration of aspatial and regional industrial policies.

The forest products sector in British Columbia provides Hayter with an unlikely example of the importance of and potential for, innovation policy. Chapter 11 suggests that innovation policy is not only relevant to the promotion of 'high technology' industries. This proposition rests upon a review of innovation theory in the context of the evolution of an industry and its relationship with product-cycle ideas. As Bradbury emphasizes in chapter 8, it is suggested that the idealized and deterministic nature of life-cycle models has encouraged an oversimplified view of industrial evolution in which maturity is follwed by inevitable decline. Hayter challenges these assumptions by presenting a case for regarding the forest products sector as a priority target for innovation policy in British Columbia. Finally, the wider implications of this case for the application of industrial and innovation policies to mature sectors are stressed.

Chapter 12 by Christy and Ironside is concerned with one of the more fashionable instruments of contemporary industrial and regional development policy – science parks. It reviews policies and programmes in this area, contending that there is no simple formula for success. Evidence from Alberta is used to test a series of hypotheses relevant to an assessment of the effectiveness of science parks as policy instruments. It is concluded that their official promotion may be regarded as an encouragement to, but certainly not a prerequisite for the development of high technology industry.

REFERENCES

Massey, D. B. and Meegan, R. A. (1982) *The anatomy of job loss: the how, why and where of employment decline* (Methuen, London).
Townsend, A. R. (1982) *The impact of recession* (Croom Helm, London).

PART I

Nature and Spatial Dimensions of Technical Change

2

Technical innovation, long cycles and regional policy

C. Freeman

The last few years have witnessed a considerable revival of interest in Kondratiev cycles (long waves). In the 1970s only a few rather isolated economists, such as Mandel (1975) and Rostow (1978) were publishing theoretical and empirical work relating to long cycles. By the late 1980s, although there was not a flood of publications, there was certainly a rapid growth in their number. The main reason for the revival of interest was the deep recession of the 1980s. If one of the tests of a satisfactory theory is predictive power, then at least in the twentieth century, long-wave theory has passed this test with flying colours. This will not, of course, satisfy those whose main interest is to demonstrate to their own satisfaction that the phenomena do not exist by the statistical measurement of earlier examples of long waves (Weinstock, 1964). In the nature of things, this group will never be satisfied, since the retrospective reconstruction of historical time series from the nineteenth century is an art rather than a science, while the number of cycles is far too small to satisfy any test of statistical significance. However, even with the imperfect evidence of the nineteenth century statistics on prices, interest rates and trade, it is worth noting that outstanding economists such as Pareto (1913) certainly took the phenomenon seriously and suggested possible explanations. Even without any resolution of the problems facing nineteenth-century historians, it is quite possible to recognize the existence of major periodic crises of structural adjustment, at least in some countries, and to endeavour to understand the nature of such crises and the problems of resuming a high and sustained growth path for the world economy (Van der Zwan, 1979).

An interesting aspect of the long debate since the earliest days has been the heterogeneous nature of those who took part. As Delbeke (1984; 1985) pointed out in his reviews of long-wave theories, there have been and still are many competing, although to some extent complementary, explanations. It may be true that Marxist economists deserve the main credit for the early development of the theory, particularly, Van Gelderen before the first World War. It is all the more welcome, therefore, that the resurgence of interest has occurred in both halves of Europe, as well as in the United States, Japan and the Third World. It was a striking feature of the Weimar Conference, organised by the International Institute for Applied Systems Analysis (IIASA) in 1985, that major contributions come from both neo-classical and a Marxist orientation.

Despite this overall pattern of diversity there is a growing interest within all these main theoretical schools in Schumpeter's (1939) perspective of technological change as a major feature, if not the prime determinant, of long-term fluctuations in the economy. For example, Kuczynski (1985) demonstrated that Marx and Engels had already made some extremely interesting observations about long-term swings in capitalist countries, and then went on to argue that major changes in technology were the type of events which would bring about big structural changes characteristic of long waves. He rediscovered Marx's comments on the influence of changes in the rate of profit on replacement investment and new investment from Volumes 2 and 3 of *Das Kapital*. Following Marx, he further argued that minor fluctuations in the rate of profit characteristic of the short-to-medium term business cycles would not be enough to shift capital out of old sectors into newly expanding fields of investment. Radical technical innovations, however, *would* be a sufficient stimulus to bring about this major realloca-tion of capital, but only as a result of a severe structural crisis of the system.

This 'rediscovery' of Marx's comments on the possibility and even probability of periodic deeper crises of structural adjustment, associated with major changes in technology, is comparable with the rediscovery of Keynes's comments (1930) on the stimulus of new technology to waves of investment. Just as more orthodox Marxist economists forgot about Marx's scattered comments on this topic, so most 'western' economist did not remember that in 1930 Keynes made the following obsevations about Schumpeter's theory:

> In the area of fixed capital it is easy to understand why fluctuations should occur in the rate of investment. Entrepreneurs are induced to embark on the production of fixed capital, or are deterred from doing so by their expectations of the profits to be made. Apart from the many minor reasons why these should fluctuate in a changing world, Professor Schumpeter's explanations of the major movements may be unreservedly accepted. (Keynes, 1930, 85).

Keynes went on to discuss and quote the central ideas of Schumpeter about entrepreneurship and the 'swarming' behaviour of firms when the pioneers had demonstrated the feasibility and the profitability of technical innovations.

This rediscovery of forgotten words of Keynes in Western Europe and of the forgotten passages of Marx in Eastern Europe is a neat demonstration of Kaldor's ironic remark that even if there are no real long waves in the economy, there are certainly long waves in the opinions and writings of economists. With very few exceptions, neither orthodox Keynesians nor orthodox Marists paid much attention to the development of long-wave theory in the 1950s and 1960s. Both groups concentrated their attention mainly on the short-term business cycles and on Keynesian-inspired attempts to smooth out these fluctuations.

SCHUMPETER'S THEORY OF NEW TECHNOLOGIES AND LONG CYCLES OF ECONOMIC GROWTH

Neither Kondratiev nor Schumpeter lived to see the great post second World War economic boom, but both the boom and the recession of the 1970s and 1980s fit well within the framework of the theories which they developed. A Schumpeterian interpretation of the fourth Kondratiev cycle would see the post 1945 boom of the 1950s and 1960s primarily as the simultaneous explosive growth of several new industries, especially vehicles, synthetic materials, petrochemicals and consumer durables. These were based on the universal availability of cheap oil and other sources of energy, and the use of mass- and flow-production techniques. It is interesting to note that neither Keynes nor most Keynesians expected that the quarter century after 1945 would be the fastest period of economic growth the world had ever experienced. Indeed, the expectations of many economists immediately after the second World War were pessimistic even about employment. This was because they often tended simply to extrapolate the problems of the 1920s and 1930s into the post-1945 world. They made no allowance for new technologies and industries and the impetus these would give to profit expectations and to new investment. When, however, the big boom did materialize, economists tended to attribute its success to the adoption of Keynesian policies. Similar forecasting errors, which make no allowance for cyclical fluctuations, were being made in the 1980s.

Schumpeter had suggested that, after a strong 'bandwagon' effect in the boom phase, with many new firms entering the rapidly expanding sectors, there would follow a period of 'competing' away of profits as the new industries matured. This would lead to stagnation and depression if a new wave of innovations and investment did not compensate. Such an explanation appears to fit the facts of the post-1945 boom rather well. It had already been remarked in the 1960s that the rate of profit was beginning to fall in several OECD countries. It fell still further in the 1970s, especially in some of the erstwhile rapid growth sectors, such as synthetic materials and consumer durables. Schumpeterians stressed four main contributory factors to this fall in the rate of profit:

1 The increased competition associated with the intensive 'swarming' and 'bandwagon' process of the high boom period and especially international trade competition. A good example of this was the entry of the oil companies into the synthetic materials industry, previously dominated by the large chemical companies.

2 The pressure on input costs associated with these same swarming processes and the rapid expansion of the economy generally. Many authors stress labour costs, but material costs and energy costs also figure prominently in some theories. Rostow's (1978) theory lays the greatest stress on the role of primary commodities.

3 The tendency after a prolonged period of expansion to approach saturation and replacement levels with respect to some new markets. There is no question of an overall satisfaction of human needs, but Pasinetti (1981) introduced the notion of a specific pattern of demand associated with each long wave. Shifts in consumer behaviour towards a new pattern do not necessarily occur as smoothly as naive utility theory might suggest.

4 The tendency to approach technical limits and scaling-up limits as the potential productivity and other gains of a new technology are more fully exploited (Wolff's Law). A good example of this is the size of oil refineries, oil tankers and petrochemical plants, which generally began to reach limits in the 1960s.

In addition to these phenomena, Mensch (1975) stresses that as a technology matures there is a tendency for entrepreneurs to move from basic product innovations to product differentiation (Scheininnovationen) and minor improvement innovations.

These explanations advanced by neo-Schumpeterian theorists for the loss of impetus in any major period of expansion offer a plausible explanation of the change in the economic climate from the late 1960s to the 1980s. Nor are they wholly inconsistent with the theories and explanations advanced by more orthodox economists. The latter, however, place far greater stress on such 'exogenous' factors as the 1973 and 1979 OPEC oil price rises, on persistent inflationary pressures within the system and on the monetary policies adopted by governments to combat these pressures. For Schumpeterians, on the other hand, the oil price increases were only one manifestation of the wider process of structural change just outlined.

When the analysis moves to the trough of the long wave more substantial differences emerge. At the time when Schumpeter first advanced his theory, Kuznets (1940) remarked that technical innovations could only sustain a long-wage expansion if they were big and important ones, since minor incremental innovations were being developed all the time. Kuznets also pointed out that Schumpeter had not explained why (when they were presumably present all the time) such key innovations seemed to occur or cluster at long intervals of about fifty years, or why entrepreneurs had a sudden burst of energy to exploit them. For those who find that a Schumpeterian approach is plausible in general, these remain important questions.

In his book on technological stalemate, Mensch (1975) presented evidence purporting to show that the bunching of 'basic' innovations was a real phenomenon, and proposed a theoretical explanation of it. Using several independent sources, he claimed that they all showed a bunching of basic innovations in three decades – the 1830s, the 1880s, and the 1930s. He explained this bunching largely in terms of two mechanisms: the effects of economic depression in stimulating and accelerating the application of radically new ideas; and a 'crowding out' mechanism that created unfavourable conditions for basic innovations during boom and 'stagflation' periods. Mensch has made an important and original contribution to this debate, but his thesis is open to criticism on both main points.

The statistical evidence for the bunching of radical or 'basic' innovations is still rather weak, and the sources Mensch used were not really satisfactory for his purpose. There *are* clusters of radical innovations in the 1880s and 1930s, but they also occurred in the 1950s and 1960s in, for example, such industries as electronics, drugs and scientific instruments. It seems more probable that acceleration effects on innovation derive from strong demand rather than depression which could inhibit risk taking. The evidence of case studies supports this view rather than the theory of depression-induced waves of innovation (Freeman, Clark and Soete, 1982).

Those who have studied case histories of actual innovations are familiar with the frequently long gestation periods and false starts. It can also be suggested that breakthroughs in basic science and technology are much more important than Mensch implies, and are more likely to bring about 'bunching' than the depression mechanism that he advocates. It can be accepted, however, that depressions may help to bring about big changes in the social and political climate (as opposed to business behaviour in firms).

These in turn may generate conditions that are more favourable in the recovery phase, both for new radical innovations and for the swarming process around older, basic innovations that may have been introduced at various times in the previous ten to thirty years. The technical interrelatedness of families of innovations must also be stressed. These are linked both by a common technological foundation as, for example, in macromolecular chemistry or microelectronics, and by common organizational and market features, as exemplified by consumer durables.

However, the important phenomenon is not a statistically observable 'bunching' of discrete innovations; it is rather the rolling of the diffusion 'bandwagon' *after* the basic innovations, and the way that the clustering of families of related innovations gives rise to the emergence of new technological systems. The original basic innovations, as such, have negligible effects on either investment, employment or Gross National Product. It is only what Schumpeter called the 'swarming' of imitators getting on the bandwagon that leads to perceptible effects in the economy as a whole. There may be quite long periods after a basic innovation is made before this bandwagon gets rolling. This swarming depends also on organizational and social innovations (Perez, 1983; 1985). This point may be illustrated by the example of the railway boom of the mid-nineteenth century. The innovation of railways was not 'induced' by the depression of the 1830s. Railway lines were in use in coal mines much earlier, and the first passenger transport railway was the Swansea train from Gower in 1809, with the first steam power services in the 1820s. Many innovations affecting track, signalling systems and rolling stock were introduced at various times from the 1820s to the 1850s. From the national economic point of view, the important phenomenon was not the precise date of these innovations but the construction of a national railway network embodying them in a new technology system. The construction of such a national network depended on changes in the capital market (the joint stock company) to mobilize capital for investment on a scale hitherto not realized, as well as on other legislative and social innovations.

A second example is the way many social changes in the 1880s and 1890s facilitated the widespread adoption of electric power in industrial production and domestic lighting in the *belle epoque* boom before the 1914–18 War. Again, widespread changes in the institutional and social framework were necessary during the Great Depression and the 1939–45 War before the boom of the 1950s and 1960s could take off. Keynesian-type reforms greatly facilitated the infrastructure investment in highways and urban development, as well as the pattern of consumer credit for widespread ownership of cars and other durable goods. The new mass- and flow-production technological style developed at this time was based on the universal availability of cheap oil, as well as electricity, and required a very different set of institutional changes from those which permitted the booms of the first three Kondratiev cycles.

These two examples are directly relevant to the questions raised by Kuznets (1940) in his original review of Schumpeter's theory of long cycles. Whether a Schumpeterian approach to the role of technical innovation can satisfy the criteria laid down by Kuznets depends upon a distinction between various types of innovation. It is necessary to distinguish not only between incremental and basic (or radical) innovations, but also to look at clusters of related innovations. A three-fold taxonomy is proposed:

1 *Incremental Innovation* This is a relatively smooth continuous process leading to steady improvement in the array of existing products and services and the ways in which they are produced. It is reflected in the official measurements of economic growth simply by changes in the input/output coefficients within the framework of an established matrix. The rate of incremental change may of course vary greatly between different industries.

2 *Radical Innovations* These are discontinuous events. They may lead to serious dislocations, economic perturbations and adjustments for the firms in a particular sector. Examples would be the introduction of television or of an entirely new material in the textile industry. As already indicated, although Mensch (1975) has maintained that the occurrence of such radical innovations is concentrated in deep depression periods, our own work suggests that their first introduction is more randomly distributed over long cycles of growth. The *diffusion* of such innovations, however, *may* take a cyclical form and may be associated with long cycles of the economy as a whole. As new products and services are diffused, they would ultimately lead to a requirement for the reclassification of an established input/output matrix by the introduction of new rows and columns.

3 *Technological Revolutions* These are the 'creative gales of destruction' which are at the heart of Schumpeter's long-wave theory. The introduction of railways or electric power are examples of such transformations. To justify the description of a 'technological revolution', a change must not only lead to the emergence of new leading branches of the economy and a whole range of new product groups, it must also have fundamental effects on many other branches of the economy by transforming their methods of production and their input cost structure. Thus a technological revolution virtually requires a new input/output matrix for a satisfactory reclassification of economic activities. Only this third category could satisfy Kuznets's requirement

for major perturbations in the system. A transformation of this kind would of course carry with it many clusters of innovations of the first and second category as previously described, which might have been introduced ten or twenty years earlier, or during the course of the diffusion itself.

Within such a framework great importance attaches to the identification and specification of the major technological transformations. For example, does nuclear technology constitute such a transformation, as some authors have suggested? Again, is biotechnology such a transformation?

THE 'INFORMATION TECHNOLOGY' PARADIGM AND REGIONAL POLICIES

In the period of structural change of the 1980s, when energy costs rose and the growth potential of the old leading sectors was partially exhausted, a new techno-economic paradigm emerged, based on the extra-ordinarily low costs of storing, processing and communicating *information*. In this perspective the structural crisis of the 1980s, like those of the 1880s and 1930s, was a prolonged period of social adaptation to a new paradigm, which affected every other branch of the economy in terms of its current and subsequent employment and skill requirements, its subsequent cost structure and its subsequent market propsects. This cluster of innovations has resulted in a drastic fall in *costs* and a counter-inflationary trend in *prices*, as well as vastly improved technical performance, both within the electronics industry and in other areas. This combination is relatively rare in the history of technology, but it means that the new technological system satisfied all the requirements for a Schumpeterian revolution in the economy.

By contrast, the new biotechnologies, although they certainly also have enormous future potential, have not yet reached the point where their macroeconomic impact could be great enough to carry the entire economy forward in the next decade or two. This illustrates the importance of the Mensch debate. It is the diffusion of the innovations of the 1950s, 1960s and 1970s, rather than the first innovations of the 1980s, which must provide the main impetus for a new economic upswing. The new biotechnologies will provide very important auxilliary growth areas and they may ultimately revolutionize agriculture, the food industry and the chemical industry; but the main elements of the new technological paradigm for the fifth Kondratiev wave cannot come from this source. Still less can nuclear technology play this role. Its applications are extremely limited and its capital cost is very high, so that any large programme would severely aggravate capital shortage problems. Its cost advantages are even now dubious and there are strong environmental, social and political arguments for limiting its diffusion, especially in the case of the fast breeder reactor.

In a research project (TEMPO) at the Science Policy Research Unit, the forty sectors of the Cambridge Growth Model were used in a systematic analysis of major innovations in the British economy. Experts from industry and technological institutions were involved in this work which contributed to the development of an Innovation Data Bank, listing a large number of innovations from 1945 to 1983 for all the main sectors of the UK economy (Clark, 1985; Guy, 1984; Soete, 1985; Freeman, 1985; Smith, 1985). The results of this research confirmed the conclusion that the new technological paradigm, which emerged in the 1960s, began to penetrate most industries and services in the 1970s. This paradigm is based on a combination of microelectronics, computerization (microprocessors in the 1970s), and tele-communications, and may be described as the 'information revolution'. The electronic-based technologies now account for *over a third* of total research and development (R&D) activities in the leading industrial countries, and for a high proportion of new manufacturing and service investment (excluding buildings).

To obtain the full benefits of the new technological system requires many changes in management attitudes and management systems, as well as other institutional and social changes. The full potential benefits can only be realized where thorough training has been carried out at all levels and the necesssary skills have been assembled, and when a comprehensive approach to the entire design and production system is adopted. There is an interesting analogy here with the diffusion of electric power in production systems in the third Kondratiev cycle. It was not until the later stages of this cycle that the full benefits of labour and capital productivity gains were realized throughout the economy and not just in a few leading sectors.

Figure 2.1 illustrates the point that, in the case of electric power, although Mensch is correct in drawing attention to a cluster of radical innovations in the 1880s, the major 'economic' effects of electrification came much later, with the growth in the share of electricity in mechanical drive for industry from 5 per cent in 1900 to 53 per cent in 1920. This was possible only after the acceptance of a major change in factory organization, from the old system based on one large steam engine driving a large number of shafts through a complex system of belts and pulleys, to a system based, first of all, on electric group drive and, later, on unit drive (i.e. one electric motor for each machine). Under the old system all the shafts and countershafts rotated continuously no matter how many machines were actually in use. A breakdown involved the whole factory.

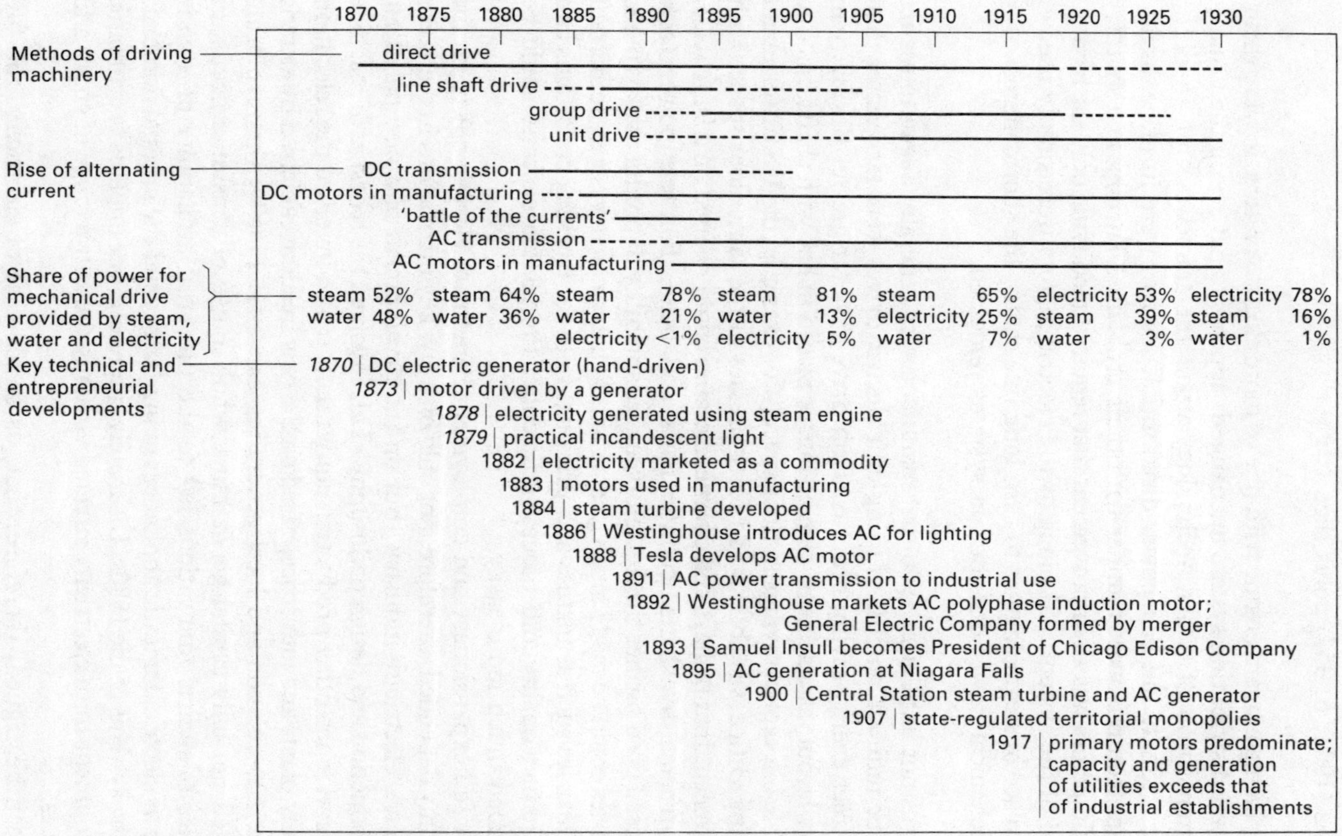

Figure 2.1 Chronology of electrification of industry
Source: Devine (1983)

Devine (1983, p. 357) notes that

> Replacing a steam engine with one or more electric motors, leaving the
> power distribution system unchanged, appears to have been the usual
> juxtaposition of a new technology upon the framework of an old
> one... Shaft and belt power distribution systems were in place, and
> manufacturers were familiar with their problems. Turning line shafts
> with motors was an improvement that required modifying only the front
> end of the system... As long as the electric motors were simply used
> in place of steam engines to turn long line shafts, the shortcomings of
> mechanical power distribution systems remained.

It was not until after 1900 that manufacturers generally began to realize
that the indirect benefits of using unit electric drives were far greater than
the direct energy-saving benefits. Unit drive gave far greater flexibility in
factory layout, as machines were no longer placed in line with shafts, making
possible big capital savings in floor space. For example, the US Government
Printing Office was able to add forty presses in the same floor space. Unit
drive meant that trolleys and overhead cranes could be used on a large scale,
unobstructed by shafts, countershafts and belts. Portable power tools
increased even further the flexibility and adaptability of production systems.
Factories could be made much cleaner and lighter, which was very important
in industries such as textiles and printing, both for working conditions and
for product quality and process efficiency. Production capacity could be
expanded much more easily.

The full expansionary and employment benefits of electric power on the
economy depended, therefore, not only on a few key innovations in the 1880s
or on an 'electricity industry', but on the development of a new paradigm
or production and design philosophy. This involved the redesign of machine
tools and much other production equipment. It also involved the relocation
of many plants and industries, based on the new freedom conferred by electric
power transmission and local generating capacity. Finally, the revolution
affected not only capital goods but a whole range of consumer goods, as
a series of radical innovations led to the universal availability of a wide
range of electric domestic appliances going far beyond the original domestic
lighting systems of the 1880s. Ultimately, therefore, the impetus to economic
development from electricity affected virtually the entire range of goods and
services.

It can be argued, therefore, that, as in the case of electricity, the full
economic and social benefits (including employment generation) of information technology depend on a similar process of social experimentation and

learning. The organizational, social and system innovations at the point of application are just as important as in the case of the electric power industry. Information technology represents a formidable challenge in terms of the need for radical institutional change. Most of our present institutions were created under, and are still geared towards, older technological paradigms. These institutions, with their self-perpetuating interest groups, represent the most formidable barriers to the rapid diffusion of information technology and the realization of its potential growth and productivity gains.

Regional development policy is one area in which institutional change will be required to take advantage of the opportunities presented by the new paradigm based on information technology (Thwaites and Oakey, 1985). The policies of the 1950s and 1960s attempted to encourage investment in less favoured regions through various financial incentives. Such measures are less appropriate in the 1980s and may, indeed, have negative effects on employ-ment. In the case of Japan, for example, regional policies are increasingly based upon: the provision of educational, technical and research infra-structures; better communications (especially telecommunications and air transport); and environmental improvement (Kuwahara, 1985). Various studies (Thwaites and Oakey, 1985) have emphasized the relevance of these factors in the UK, and Morgan and Sayer (1985) have stressed the importance for occupational structures of switching from 'branch–plant' investment to R&D and technical activities.

New types of regional policy may be facilitated by information technology. Several advantages have been identified by Soete (Freeman and Soete, 1985). First, the dramatic potential to separate actual production from control, administration, design, management and marketing. This offers substantial scope for the geographical relocation of firms and businesses. In a wide variety of occupations it opens up the possibility of work at home. Second, the realization of this potential depends crucially on the widespread availability of a fully interactive (two-way) communications network. Third, the de-centralization potential of information technology will allow in a far more effective way for the creation of local, quasi-autonomous growth poles. Just as in the case of economies of scale, information technology questions the very basis of the agglomeration effects generally associated with growth concentration. Fourth, and following to some extent from the previous points, information technology could ultimately lead to the reduced importance of daily commuter–type transport services. Despite these tendencies towards decentralization, Goddard et al. (1985) pointed out that information techno-logy can also be used to reinforce centralizing power to the disadvantage of the less developed regions. Brotchie et al. (1985) speculated on the longer-term implication for urban activities, without, however, any clear resolution of these tendencies and counter-tendencies. More precise forecasts may be impossible since the eventual outcome is a matter of social and political choices and conflicts.

CONCLUSIONS

These examples of structural changes and policy changes associated with the widespread introduction of new technology are relatively straightforward compared with the fundamental problem of restructuring the world economy and especially of putting north–south relationships on a new basis. It is here that the uneven development of the world economy is most evident and raises the most intractable problems. National policies for regional development in OECD countries are far from perfect, but international policies are weaker still.

The establishment of the IMF and the World Bank and the general adoption of expansionary Keynesian policies after the 1939–45 War did temporarily create a fairly stable framework for the growth of the world economy. Large-scale public and private investment in Third World countries also encouraged the efforts of many of these countries to catch up with leading OECD countries. But in the 1980s, these rather favourable conditions for the poorer countries to improve their situation ceased to exist. The burden of debt repayment became so great that it constituted a major source of economic and political instability, especially in Latin America and Africa. There was a net *outflow* of funds from many Third World countries, and a net *inflow* into one of the richest countries, the United States, which financed the enormous deficit in US foreign trade. This deficit was also related to the relative failure of the US economy to adapt to the new paradigm outside the military area, by comparison with the much more successful adaptation by Japan which resulted in the high Japanese surplus. This situation cannot be sustained and points to the need for fundamental structural change in the international machinery for adjusting international payments and capital transfers as well as to major policy changes within OECD countries. The real resources available to the IMF and the World Bank are now grossly inadequate compared with the vastly increased volume of international trade and investment.

The Inter-American Development Bank (September 1985) highlighted these critical problems confronting the world economy and warned that the IMF measures to deal with the crisis were short-term palliatives, not long-term solutions. Albert Fishlow in his contribution to the survey points out that Latin America faces a burden of debt service repayments double the level of reparations that Germany found impossible after the 1939–45 War. It will be remembered that Keynes made a devastating critique of the German reparations arrangements and pointed out that they would result in world economic collapse and would probably lead to another war. In Latin America only a recovery of productive investment and technical innovation could sustain the growth needed to finance even a much lower level of debt repayment.

The Third World countries are also experiencing difficulties in developing the new information technology industries to sustain their competitive power. However, as Perez (1985) has pointed out, the new technologies do actually offer some major advantages to Third World countries, provided they modify their trade, industrial and technology policies. In sectors such as software engineering and telecommunications equipment, there are big new opportunities for developing countries, both in local applications of information technology and in world markets.

However, these 'catching up' efforts of Third World countries require some resolution of the basic structural problem confronting the entire world economy. This imples new measures to facilitate the international transfer of technology as well as a resolution of the debt problem.

But as in the 1880s and 1930s, in confronting a major structural crisis of adaptation, there are many alternative courses and the outcome will depend on the balance of social and political forces. These are strong pressures to restore the general rate of profit, not so much by technical change and imaginative social improvements, but by a reactionary combination of military expenditures, increased protectionism, and reduction in the power of trade unions, as well as in centralizing policies at the expense of autonomous regional developments. This creates an urgent need to develop imaginative new initiatives for the growth of the world economy, involving the widespread use of the new paradigm. The regional dimensions of such policies are clearly of exceptional importance.

ACKNOWLEDGEMENT

I am grateful to the ESRC whose support for the SPRU research programme in Technical Change and Employment (TEMPO) made this work possible. It is based in part on lectures given to the Siemens Foundation in Munich and the Kikawada Foundation in Japan.

REFERENCES

Brotchie, J., Newton, P., Hall, P. and Nijkamp P. (Eds.), (1985) *The Future of Urban Form: The Impact of New Technology*, Croom Helm, London.

Clark, I. A. (Ed.), (1985) *Technological Trends and Employment*, Vol. 2, *Basic Process Industries*, Gower, Aldershot.

Delbeke, J. (1984) Recent Long Wave Theories: A Critical Survey. In Freeman, C. (Ed.) *Long Waves in the World Economy*. Frances Pinter, London.

Delbeke, J. (1985) Long Wave Research. The State of the Art, 1983, in IIASA *Proceedings of the Siena Florence Meeting*, Laxenburg.

Devine, W. W. (1983) From Shafts to Wires: Historical Perspectives on Electrification, *Journal of Economic History*, 43, 347–372.

Dupriez, L. H. (1947) *Des Mouvements Economiques Generaux*, Louvain.

Freeman, C., Clark J. and Soete, L. L. G. (1982) *Unemployment and Technical Innovation: A Study of Long Waves in Economic Development*, Frances Pinter, London.

Freeman, C. (Ed.), (1984) *Long Waves in the World Economy*, Frances Pinter, London.

Freeman, C. and Soete, L. L. G. (1985) *Information Technology and Employment*, IBM, Brussels.

Freeman, C. (Ed.), (1985) *Technological Trends and Employment*, Vol. 4, *Engineering and Vehicles*, Gower, Aldershot.

Freeman, C. and Soete, L. L. G. (1986) *Technical Change and Full Employment*, Blackwell, Oxford.

Goddard, J. et. al., in Thwaites, A. T. and Oakey, R. P. (1985) *The Regional Economic Impact of Technological Change*, Frances Pinter, London.

Guy, K. (Ed.) (1984) *Technological Trends and Employment*, Vol. 1, *Basic Consumer Goods*, Gower, Aldershot.

Inter-American Development Bank, (1985) *Annual Survey*, Washington, September 1985.

International Institute of Applied Systems Analysis (IIASA) (1985) Collaboration Paper: Proceedings of the Siena/Florence Meeting, Laxenburg.

Keynes, J. M. (1930) *Treatise on Money*, MacMillan, London.

Kleinknecht, A. (1981) Observations on the Schumpeterian Swarming of Innovations, *Futures* 13, 293–307.

Kuczynski, T. (1985) *Marx and Engels on Long Waves*, Institute of Economic History, Berlin, paper given to IIASA Conference, Weimar.

Kuwahara, Y. (1985) Creating New Jobs in High Technology Industries, paper presented at OECD Conference on Employment Growth in the Context of Structural Change, Paris, OECD.

Kuznets, S. (1940) Schumpeter's Business Cycles, *American Economic Review* 30, 257–271.

Maier, H. (1985) *Basic Innovations and the Next Long Wave of Productivity Growth: Socio-Economic Implications and Consequences*, IIASA, May 1985 (paper for the Weimar Conference, June 1985).

Innovation and long cycles 25

Mandel, E. (1975) Spätkapitalismus, Frankfurt.

Mandel, E. (1984) Explaining Long Waves of Capitalist Development. In Freeman, C. Ed.) *Long Waves in the World Economy*, Frances Pinter, London.

Mensch, G. (1975) *Das technologische Patt*, Umschau, Frankfurt.

Menshikov, S. and Klinenko, L. (1985) On Long Waves in the Economy in IIASA, *Proceedings of the Siena–Florence Meeting*.

Morgan, K. and Sayer, A. (1985) The International Electronics Industry and Regional Development in Britain, Sussex Working Papers, 1983, and paper presented to IBG, CAG Symposium, Swansea, August 1985.

Pareto, V. (1983) Alcuni Belazioni fra lo Stato Socialle e le Variazoni della Prosperita Economica, *Rivista Italiana di Sociologia*, September–December, 501–48.

Pasinetti, L. L. (1981) *Structural Change and Economic Growth: A Theoretical Essay on the Dynamics of the Wealth of Nations*, Cambridge University Press.

Perez, C. (1983) Structural Change and the Assimilation of new Technologies in the Economic and Social System, *Futures* 15, 357–375.

Perez, C. (1985) Microelectronics, Long Waves and World Structural Change, *World Development* 13, 441–463.

Perez, C. (1985) Towards a Comprehensive Theory of Long Waves in IIASA, Proceedings of the Siena–Florence Meeting.

Rostow, W. W. (1978) *The World Economy: History and Prospects*, University of Texas Press, Austin and Macmillan, London.

Smith, A. (Ed.) (1985) *Technological Trends and Employment*, Vol. 5, *Commercial Service Industries*, Gower, Aldershot.

Schumpeter, J. (1939) *Business Cycles*, McGraw-Hill, New York.

Soete, L. L. G. (Ed.) (1985) *Technological Trends and Employment*, Vol. 3, *Electronics and Communications*, Gower, Aldershot.

Thwaites, A. P. and Oakey, R. P. (1985) *The Regional Economic Impact of Technological Change*, Frances Pinter, London.

Weinstock, U. (1964) *Das Problem der Kondratievzyklen*, Berlin.

van der Zwan, A. (1979) On the Assessment of the Kondratiev Cycle and Related Issues, Centre for Research in Business Economics, Erasmus University, Rotterdam, 1979, mimeo.

3

Innovation, diffusion and technical change: a case study

R. McArthur

An analysis of the factors affecting the rate and timing of the diffusion of a piece of graphic arts machinery, the colour scanner, raises a whole variety of questions about the way that innovation, diffusion and technical change are interpreted. Can technical innovation and diffusion be treated separately? Which of these two has the greater economic impact? What is the importance of technical innovation in the pace and timing of technical change? What role do innovations in market relations and innovations in the social relations of production play? In providing some tentative answers to these questions, it is necessary to draw upon a broader body of work than that concerned with innovation diffusion alone. The first part of the chapter describes the pervasiveness of 'heroic' views of innovation and some of the reactions to such views; the second part considers the treatment of innovation diffusion in three distinct strands of literature; the third section presents an analysis of the diffusion of the colour scanner; finally, some of the implications of this analysis for attempts to increase the rate of diffusion and for spatial policy in a broader sense are explored.

INNOVATION, DIFFUSION AND TECHNICAL CHANGE

There are innumerable folk songs and popular ballads about the steam engine. There are none to my knowledge on the problem of the precision boring of cylinders which was a necessary step before a reliable steam engine could be constructed. Neither do I know of any of the advantages of distributed electric drives compared to belts and pulleys (Devine, 1983, referred to by Freeman in chapter 2). Similarly, there are pop songs in which computers figure but none that I know of about the problems of machine compatibility or interfacing – despite the great scope for analogies! In other words, the major innovation has an imaginative appeal which serves to heighten its apparent importance compared to the other small developments which are equally crucial to its success. Some are enabling techniques such as the cylinder boring machine; others, like distributed drive, are factors which affect the uses to which the device is put and the way in which it is developed, that is, factors affecting diffusion.

In numerous articles Rosenberg has pointed out this bias towards the 'heroic invention' and the 'heroic innovator' at the expense of the crucial enabling factors which affect the rate and timing of technical change (Rosenberg, 1976). On a different front, historians have been developing alternative explanations of industrialization to the 'Prometheus Unbound' of Mantoux and Landes' English Industrial Revolution sweeping all before him. They point to the diversity of forms of industrialization before the Industrial Revolution, the small and progressive artisanal innovations and the evolution of different patterns of capital accumulation (Berg, Hudson and Sonenscher, 1983). Equally they point to the different paths to the twentieth century that industrialization followed in different countries, none of which can be considered as inferior to the English model of concentrated capital (Sabel and Zeitlin, 1985; O'Brien and Keyder, 1978). Indeed there are indications from literature on France and Italy that more dispersed, less urbanized industrial systems have more stability and growth potential (Fua, 1985; Bagnasco, 1985; Brusco, 1986; Saglio, et. al., 1984). Prometheus is paying the price for his precocity.

Just at the time, then, when the possibility of a new industrial revolution or a new Kondratiev wave is being considered, the primacy of major innovations in the process is being challenged. This does not involve a contradiction, but rather the need to qualify the role of innovation by recognizing the more gradual pace of the incorporation of technical innovations into the industrial fabric, the importance of economic and social conditions in the timing and pace of uptake of innovations and, as will be shown by an example, the changes that these prompt in innovations themselves. It will be suggested then that the move towards a more connected and flexible view of the role of innovation in economic development is hampered by heroic views of innovation, often only implicitly held, and by the way that the study of the diffusion of innovations has developed.

INNOVATION

Some of the problems that have arisen in the analysis of technical change from an overemphasis upon innovation can be conveniently related to two debates: one on the role of innovation in economic development and changes in resource productivity and the other on the nature of 'high-technology' industries.

Problems in the first debate are well illustrated by Mensch's account of how innovations overcome the depression of the Kondratiev wave (Mensch, 1975). As Freeman has described in chapter 2, Mensch considered that basic innovations were bunched in the depression of the long wave as this was when conditions were most favourable to their appearance. Once they appeared and the possibilities were perceived for making greater profits than with existing investment, producers and investors 'swarmed' to take advantage. Following Freeman, it can be demonstrated that these basic innovations are, in fact, spread over a longer period and, as with any innovation, they go through a more or less protracted phase of experimentation and development. The 'bandwagon' effect is associated with diffusion: with the development, modification, adoption and application of the innovations in a whole variety of ways.

It would be perverse to argue (restrospectively) that the integrated circuit was not a basic innovation whilst the Intel CMOS microprocessor was, on the basis that the former appeared in the early 1960s during an upswing, whilst the latter appeared in 1972 during downswing and depression. They are both part of integrated circuit development which has been a gradual and drawn out process, the economic impact of which depends upon the cumulative development, modification and improvement which accompanies diffusion (Dosi, 1981). Neither could the impact be said to have been sufficient so far to produce a large enough bandwagon effect to speed up the rate of growth in the whole economy. The process of innovation development is more gradual and tentative than Mensch suggests, and is punctuated by incremental improvements which are just as important as the 'original' innovation in terms of their economic effects, if not more so. Mensch has usefully drawn to our attention the links between innovations and cyclical economic activity but has, it would seem, described those links in the 'heroic' fashion described earlier rather than recognizing the significance of development and diffusion – in which depression may play an important part.

A second example of the overemphasis placed upon innovation is provided by the role attributed to Research and Development (R&D) in uneven development. Most innovation in modern economies can be traced, so the argument runs, to R&D programmes (Freeman, 1982). Variations in productivity growth rates across sectors or between regions within the same sector can be positively related to some measure of R&D intensity, although which of these is most appropriate is a matter of controversy (Kendrick, 1973; Nelson and Winter, 1975; Nelson and Winter, 1977; Scherer, 1982). From there it is but a short step to argue that reallocating the Promethean fire between sectors and between regions is the only long-term way of removing disparities. Recognition is certainly made of the problems involved in so doing, in terms of attracting high-status activities or science-rich activities to peripheral areas for example, but the basic building block for prosperity is seen to be the level of R&D and the problem is perceived as being how to increase it (Howells, 1984).

Contrast this with Nelson and Winter's suggestion that different sectors and different industrial structures have varying capacities, which differ over time in their ability to carry out useful R&D (Nelson and Winter, 1977) Pumping R&D may have no more than a marginal effect since differential productivity levels may turn more crucially upon a sector's technological base or the flexibility of its industrial organization. That a sector's technological bases change historically and that some develop more rapidly than others has been demonstrated by Freeman and his co-workers at SPRU. Indeed, this fits in with the idea of shifts in key technological systems (Freeman et. al., 1982). The effect of flexibility of industrial organization is in danger of becoming trivialized in current discussions of flexible manufac-turing and flexible working. It is worth commenting at this point, therefore, on the very different forms of industrial organization that have been emerging in Italy and France, particularly since the early 1970s, which draw on a very different kind of social organization. These are small, highly innovative firms producing for national and international markets, based in small town areas (e.g. Emilia Romagna, Veneto) drawing upon stable social networks and local collaborators and subcontractors. In this way, they produce, using modern machine tools in particular, a whole range of short-run components and finished goods. Small size, the ease of opening and closure (no stigma is attached to failure), social mobility and close contacts between owner and worker all make for a flexible and responsive system of production. (The only account in English that the author knows of is Brusco (1986); the most complete accounts are in Bagnasco (1985) and Bagnasco and Triglia (1985).

Much is still unclear about these industrial systems but it seems to be important to signal that the path towards productivity growth, innovation and, indeed, regional development may turn as much upon industrial organization in the broad sense as on the supply of innovations. This is a view which is impeded by views which take an instrumental view of innovation such as those described above. The second debate which shows a similar overemphasis on innovation is that on identifying 'high-technology' industries.

Although deciding which industries are 'high technology' is a matter of debate and uncertainty, there seems to be some consensus that such industries are 'science- or knowledge-based' and have high levels of innovation

(Langridge, 1984; McQuaid, 1984). Various criteria are then used to identify these 'high-technology' industries, such as R&D expenditure per employee, sectors with the largest growth rate in patents or proportion of technical employees. Whatever the measures used, the reasoning is largely the same: these are the industries which, by virtue of high innovation rates (and hence growth rates), will lead the way out of depression and, therefore, require supporting or, in the case of regions, capturing. This analysis may or may not be explicitly tied to a neo-Schumpetarian or Kondratiev wave framework. If it is, added urgency is attached to capturing the industries of the next Kondratiev wave (Hall, 1982; Rothwell, 1982). However, this kind of approach seems to miss the significance of identifying the role these technologies play in the economic structure. On the basis of measures of innovation, it is possible to conflate technologies such as biotechnologies which are still largely in experimental and development stages, with those such as the production of computer process control technology which are playing an important role in current changes in industrial structure and resource productivity across a wide range of industries. Both might have similar numbers of innovations, R&D levels or technical employees but they have very different economic impacts. Surely the level of adoption, the kinds of uses and the kinds of effects are the distinguishing features – features which cannot be left unstated nor read off from surrogate measures of innovation.[1] Such diagnostic characteristics imply that the timing of technical change turns to a very considerable extent upon the rate of adoption of innovations and perhaps upon the links between innovation producers and users. If the overemphasis on innovation is to be corrected, it is necessary, therefore, to have as a counterweight a good understanding of the diffusion process and the range of factors which affect the rate and timing of diffusion.

DIFFUSION

Nathan Rosenberg wrote in 1972:

> ...it is a striking historiographical fact that the serious study of the diffusion of new techniques is an activity no more than fifteen years old...Even today, if we focus upon the most critical events of the industrial revolution...our ignorance of the rate at which new techniques were adopted, and the factors accounting for these rates is, if not total, certainly no cause for professional self-congratulation (Rosenberg 1976, p. 89).

Although the study period has doubled and considerably more work has been done on innovation diffusion since 1972, it could be argued that our understanding of diffusion still does not provide a sufficient counterweight to the heroic views of innovation previously described. Part of the reason for this is that differences in methods of analysing innovation have prevented much of the knowledge gained from being integrated. In particular, the methods and assumptions behind studies of innovation diffusion in the economics tradition are not easily compatible with studies in economic history and explanations of the timing of technological change associated with the long-wave debate.

The analysis of innovation diffusion in economics has been developed from epidemic theory. In essence, the rate of adoption (infection) by the population of potential adopters (those at risk) is seen to depend upon the characteristics of the innovation (infectiousness) in terms of investment required or availability, for example, and upon the features of adopters such as their size, access to information or progressiveness. The rate of adoption is described in terms of its fit to some kind of S-curve to see how the various characteristics of both the innovation and the adopters affect the rate. Recommendations are typically made as to how the rate of adoption can be increased by altering various characteristics of the innovation (e.g. investment required) or by modifying features of the adopters (e.g. their access to information). Classic accounts are those by Mansfield (1961) and those contained in Nabseth and Ray (1974).

Considerable sophistication has been added to this model by modelling the effect of supplier strategies or changes in firm sizes over the adoption period (Stoneman, 1976; 1983). Furthermore, the use of more sophisticated modelling techniques such as probit analysis has allowed some earlier restrictive assumptions to be relaxed. Davies (1979) developed one such model for the diffusion of process innovations which has had some application in the regional development literature (Alderman, 1985). Davies argued for a behavioural view of firm strategy in which the decision to adopt can be related to the perception by a firm of such things as the innovation's profitability, the payback period required and whether adoption by competitors increases or decreases the need to adopt. In addition, Davies views firm behaviour and the characteristics of the innovation as resulting in two basic diffusion curves, one approximating to a cumulative normal curve and the other to a cumulative log-normal curve, not dissimilar to the logistic curve of earlier studies. The first curve represents the case where the innovation is simple, cheap and the advantages of adoption are quickly demonstrated: this results in a rapid adoption rate among potential adopters. The second curve represents innovations that are more complex and have more teething problems: hence the advantages of adoption are not initially so evident. Unfortunately, due to the limitations of multivariate analysis, the behavioural aspects of his theory have to be reduced to one single proposition: that firm size is one key variable which reflects all the others in some measure. Nevertheless, he is able to demonstrate that firm sizes in an industry do affect the rate of adoption and the time taken for one hundred per cent adoption to be reached.

Whatever the variations in these approaches, they share a set of critical common features: the belief that a set of potential adopters can be identified; that changes in the innovation may affect the rate of take-up of the innovation or make it easier for smaller firms to adopt, but essentially the group of potential adopters remains the same; that the diffusion curve has a prescriptive and analytical value which allows it to be used for forecasting and policy making.

Consideration of innovation diffusion in economic history has in some cases been close to the analyses described above (e.g. David, 1975), but there is also a more descriptive body of literature dealing with the adoption of innovations. The analysis in this work is concerned with accounting for variations in diffusion rates in terms of variations in institutional economic and social characteristics (Fogel and Engermann, 1971; Rosenberg 1976). For example, the early diffusion of the steam engine has been related to the problems of obtaining skilled workmen, the role of capital institutions in providing finance and, not least, learning what tasks were appropriate and economic for steam power (Tann, 1981; von Tunzelmann, 1978). The other dimension to historical diffusion studies is that developed by Rosenberg in which the importance of developments to an innovation during the course of its diffusion is recognized. Indeed, he demonstrates how modifications to innovations as diverse as machine tools and steam engines resulted in major changes in the groups of adopters (Rosenberg, 1976, chapters 1, 10 and 11).

Turning to explanations of the role of technical change in the timing of long waves, it has become increasingly clear through the work of Freeman and his co-workers, Perez and others, that the constraints upon the application, development and diffusion of new technologies turn upon the imbalances and destabilization of both supply and demand that may be associated with both the limited introduction of major new technologies and the rigidities of the existing organization of production (Freeman et. al., 1982; Perez 1983; Pasinetti, 1981). As Freeman suggests in chapter 1 and elsewhere there has been a failure to realize the technological potential for productivity increases that have been available in the 1970s and 1980s. He also indicates that such features of the institutional and social framework as 'the education and training system, the industrial relations systems, managerial and corporate structures, prevailing management styles, the capital markets and financial systems. . .all act as constraining factors (Freeman, 1984).

It is often difficult to reconcile the views of diffusion associated with the epidemiological modelling of economists, the more descriptive approaches of economic historians and the explanations related to the long-wave debate. It could be argued, though, that the problem is simply one of scale and that the collection of enough variables representing, say, skill levels, access to capital, managerial techniques and products manufactured would allow these broader concepts of diffusion to be modelled using mathematical techniques. Indeed, it may be said that a series of studies based at the Centre for Urban and Regional Development Studies (CURDS) have done just this.

Whatever one may think of the validity of reducing such complex features to modellable variables, some basic problems remain. If the innovation changes in the course of diffusion and as a result potential adopters change – as Rosenberg suggests – it becomes very difficult to describe a population of potential adopters. Furthermore, if the changes in the innovation, changes in adopters and changes in, say, product and factor markets are mutually dependent, multivariate analysis becomes well nigh impossible. This is, though, the import of a critical review of innovation diffusion by Gold (1981) who considered that there was no single 'effect' of adoption in terms of payback or profitability, but rather a 'changing pattern of effects' as uptake of the innovation by the population of adopters affected the supply of factors of production and the nature of product markets.

Although these problems make detailed mathematical analyses impossible at present, it may well be possible to describe the diffusion of an innovation in a rigorous way taking these features into account. Demonstration of this with an example of a current widely diffusing technology would provide a means of linking up individual case studies with more general views of the rate and timing of technical change.

DIFFUSION OF THE COLOUR SCANNER IN THE UK[2]

The colour scanner is a process innovation which is used in the reproduction of colour illustrations for printing. Although not as well known as, say, CAD or CNC machine tools, it shares with them the important characteristic of reducing the capital cost of a basic input in production, in this case all forms of colour illustration for printed products. It does so using a combination of computer, laser and photochemical technologies. Its use has allowed the cost of the basic product, colour transparency sets, to remain constant in monetary terms since the beginning of the 1970s (around £12–15 per set): that is, to decline massively in cost in real terms. The cost of the equipment has also remained more or less constant in real terms despite very great improvements in performance. As such, it has properties analogous to those technologies which Freeman describes in chapter 2 as being capable of fuelling an upswing, although its economic effects are on a much smaller scale. By examining the factors affecting the rate and timing of its diffusion, some insight can be gained into the factors affecting the diffusion of new technologies in general.

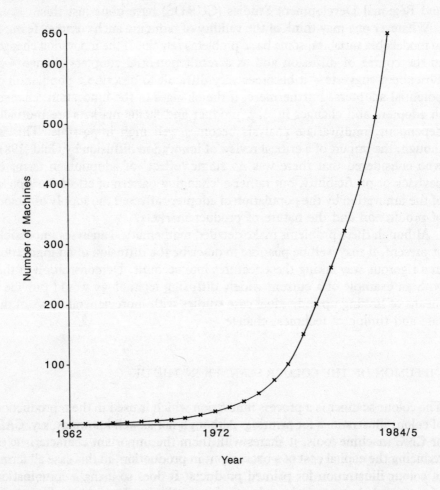

Figure 3.1 Cumulative annual sales of colour scanners in the United Kingdom 1962–1984/5

To start in a conventional way, the adoption curve, that is the cumulative sales curve, can be plotted over time. This is shown in figure 3.1. At first sight this looks like some kind of S-curve that has not yet begun to tail off. Indeed, using a simple least-squares curve fitting exercise, the diffusion curve fits Davies' long-cumulative curve for innovations with teething problems whose application is not initially evident, remarkably well. If adopters (table 3.1) are examined, it looks superficially as though large establishments adopted first, followed by smaller establishments. As table 3.1 shows, establishments with more than two hundred employees were responsible for 85 per cent of early adoptions before gradually falling to around 10 per cent for later periods. All this seems to fit in quite well with standard diffusion analysis. If the technical changes that have taken place in the colour scanner are examined, however, a different picture begins to emerge.

Table **3.1** Characteristics of early scanner users; employment and main activity, 1962–1974

Numbers employed at time of adoption	a	b	c	d	e	f	g	h	i	j	% in each size band
200+	2	3	6								85
100–199		1									7.5
50–99	1										7.5
25–49											
0–24											
% in each class	23	31	46								

Number of establishments	9	a	scanning only
Number of installations	13	b	graphic arts repro house
		c	printing offset litho
Machines: Scanatron (Crosfield)		d	printing and packaging, gravure
Colourgraph (Hell)		e	all processes, printing
Autotron (HPL)		f	all processes, packaging
Diascan (Crosfield)		g	newspapers
		h	newspapers and periodicals
		i	specialised printing
		j	other, nonprinting

Table 3.2 shows in some detail the changes that have taken place in the performance and characteristics of the colour scanner from 1962 to 1985. This involved the transformation of the scanner from a limited tool for colour correction to a machine capable of colour reproduction and correction, and then to one able to carry out electronic page make up as well. What figure 3.1 now seems to suggest is that the early period of slow diffusion

Table 3.2 Some major changes in the performance and characteristics of colour scanners used in the United Kingdom.

Year	Minimum time to produce a colour set	Changing characteristics	Company introducing changes and first production model	Implications
1962	30 min (colour correction only)	Colour creation of previously made sets only		Final size sets still have to be made using a process camera
1967	30 min	Enlarging from a colour transparency	Hell, PD1	Allows sets to be made from the original: a step towards the scanner becoming a full colour reproduction tool
		Direct screening	Crosfield Magnascan 450 Hell Chromographs 285, 286, 287	all colour process work carried out on one machine
		Four sets output at a time	PD1 MRD 1969, Crosfield 1974	Increases productivity on small format work
1972				The scanner is now a self-contained unit for colour reproduction and correction
1974	15 min		Crosfield Magnascan 550	Potential for integrating the scanner directly into electronic systems; use of push-button controls
1975		Laser scanning	Hell 300ER	More accurate picture definition
1977		Addition of page make-up facilities	Crosfield Magnascan 570	Begun to address the productivity bottleneck caused by the need to make up pages by hand
	11 min	Electronic Dot Generation	?	No need for contact screens; better dot control allows more accurate colour reproduction on a wider variety of substrates
1980		Introduction of electronic page make-up workstations which include colour correction and image editing as well as page make-up	Scitex	The versatility of the scanning system is dramatically increased as creative use now becomes possible
1982		Addition of microcomputer facilitating job queuing, storage and automatic output	Crosfield 640/5	Increases the flexibility of work organization

was associated with a machine of limited applicability, whilst the rapid increase in diffusion in the early 1970s was related to the technical changes which resulted in the colour scanner becoming a versatile colour reproduction and correction device. In other words, changes in the rate of diffusion seem to have been influenced to a considerable extent by changes made to the innovation during the course of diffusion.

Returning to table 3.1, it can now be seen that the adopters of the colour scanner have changed in ways that might be related to the changes in the scanner. Thus, the first users were large printing houses or graphic arts producers (repro houses) without exception; the scanners they used were highly limited and restricted to certain kinds of work (table 3.3). Adopters, therefore, needed a high throughput or, alternatively, had to be willing to experiment with the equipment, believing in its future potential. This latter was certainly the case with some of the early adopters and they played a crucial role in the early development of the technology. Increases in scanner versatility were accompanied by a reduction in the size of adopting establishments and a spread of the main activities of adopters, including adopters whose only activity was colour scanning (tables 3.3–3.6). The most recent improvement, electronic make up (table 3.6), is associated with a slightly greater number of multiplant enterprises but, despite its greater cost, not with larger establishments (*c*.£250 000 000 compared to £60 000–120 000 for a basic scanner).

Table 3.3 Users of pre-digital Crosfield Magnascans and early Hell Chromograph users, 1969–1978

Numbers employed at time of adoption	a	b	c	d	e	f	g	h	i	j	% in each size band
200+		2	5	2	1		1	4	1		24
100–199		1	1								3
50–99		7	2	1							15
25–49	6	9	4								29
0–24	7	10									28
% in each class	19	43	18	5	2		2	6	2	2	

Number of establishments 54
Number of installations   67

Machines: Magnascan 450, 460
         Hell DC 300, 300L, 300ER

Table 3.4 Use of early digital scanners and Hell Chromograph with no laser imaging or electronic screening, 1975–1982

Numbers employed at time of adoption	a	b	c	d	e	f	g	h	i	j	% in each size band
200+		1	5	4		1		8	2	2	10
100–199		3	1								2
50–99		8	5								6
25–49	7	28	9								19
0–24	63	80	1								63
% in each class	31	53	9	2		1		4	2	1	

Number of establishments   185
Number of installations   229

Machines: Crosfield 510, 515, 520, 500
          Hell 299, 299L, DC300

Ownership by multiplant enterprises

Activity of main enterprise	Size of scanning establishment	
all printing processes	1–24	3
packaging and papermaking	25–49	6
non-printing activity	50–99	
	100–199	1
	200+	1

How are these changes related? Did changes in the scanner permit a new group of adopters to emerge or did adopters themselves bring pressure to bear upon the producing companies? Were size and specialization of adopters exogenous or in some way related to the technical changes that occurred? Has the 'changing pattern of effects' on the production environment affected the possibilities for adoption as well? Questions such as these are not often posed in diffusion studies but they are necessary components of diffusion if the way diffusion affects both rate and timing of technical change and, indeed, innovation itself are to be understood.

ROLE OF EQUIPMENT PRODUCER STRATEGIES

Most colour scanners are produced by one of two firms, one British, the other German. They were joined in 1979 by an Israeli firm making the

Table 3.5 Digital scanners and scanners with laser imaging, electronic screening and system compatibility, 1981–1985

Numbers employed at time of adoption	a	b	c	d	e	f	g	h	i	j	% in each size band
200+		2	6	2			3	2		1	10
100–199		4	5	1							6
50–99		4									3
25–49	23	20	2								27
0–24	36	54									54
% in each class	36	51	8	2			2	1		1	

Number of establishments 178
Number of installations 220

Machines: Crosfield 530, 540/630, 640, 610/25/35/45
Hell 399, 399ER, DC350, DC350ER, CP340, CP341ER

Ownership by multiplant enterprises

Activity of main enterprise	Size of scanning establishment				
	1–24	25–49	50–99	100–199	200+
printing	3	13	–	1	3
non-printing	–	–	–	–	1

percentage of users unidentified 25%

electronic page make up component and, in 1984, by a Japanese firm. The two main firms were responsible for over 90 per cent of all installations worldwide. One might thus expect their strategies to have affected the development of the scanner. Metcalfe (1981) argued that any bandwagon effect in the process of economic development depended upon the adequate supply of process equipment. This equipment had not only to be profitable to use but also profitable to produce. Furthermore, the producer had to install sufficient productive capacity before any rapid expansion of output could occur. There is no evidence that the early scanners were not profitable to produce, so this may be rejected as a reason for the slow initial growth rate. However, the general insight is a valuable one – the problem is that it needs elaborating since the best strategy for achieving profitability is not likely to be clear in situations of uncertainty and rapid technical change.

First, there has been no one obvious path of logical technological development that scanner producers may follow. One company devoted its resources

Table 3.6 Users of electronic page make-up systems, 1980–1985

Numbers employed at time of adoption	a	b	c	d	e	f	g	h	i	j	% in each size band
200+		1	1				1	1			7
100–199		3									5
50–99		3									5
25–49		12	13	1							45
0–24	13	10									38
% in each class	43	52	3				2	2			

Number of establishments   63
Number of installations   69

Machines: Dainippon 5G2000
Hell Chromacom and Compact
Crosfield 820, 840 and 860
Scitex Vista Imager and Response

Number owned by multiplant enterprise

Main activity	Size of scanner establishment				
	1–24	25–49	50–99	100–199	200+
printing	–	8	–	–	2
printing and packaging	–	–	2	–	–
non-printing	–	–	–	–	–

known no. of disadoptions of systems 4

percentage of users unidentified 16%

in the early 1970s to the development of a digital scanner and followed a course of attempting to capitalize upon the advantages of an all-digital system. At a similar time its main competitor concentrated upon improvements to an analogue scanner by introducing laser scanning and electronic screening with a view to consolidating a reputation for speed and versatility. Thus, colour scanner producers have exercised options and choices about the direction of scanner development and have not followed a technologically predetermined course. At the outset the research and development strategy of producers is constrained by the technologies available to them, but this still provides latitude for resource allocation in R&D and for divergences in development from company to company which will affect the timing of the introduction of technical changes and the direction in which technical change proceeds.

Second, the suppliers' marketing strategies will also affect the profitability of production, the timing of technical change and the rate of diffusion. At its simplest, the rate of diffusion may be constrained by a lack of suppliers' productive capacity to meet demand. This occurred in the mid- to late-1970s with scanner producers, much as Metcalfe suggested. More subtly, the rate of diffusion may be affected by the producer's competitive behaviour. For example, one scanner producer has pursued a strategy of maximizing the installed base among client firms through financing deals, in the belief that as a result it will benefit from an assured market for upgrades and replacements over the long term. Likewise, all the producers are taking competitive measures to increase their installations of the additional electronic page-make-up components, despite no clear evidence of their current profitability for suppliers or users. This is in part related to a desire to open up a new market and in part because they see the expansion of the stand-alone scanner market drawing to a close. Thus market control may be a more potent influence on behaviour than profitability alone.

Third, the equipment producers' strategies are affected by their customers. In the case of scanners, much of the development work has been carried out with certain larger key users who have been responsible for ironing out equipment problems. This intertwining of producer and users is probably the norm (Pavitt, 1984). Indeed, it may be difficult to say exactly where the innovation has come from (von Hippel, 1978). Although this interaction may allow larger users to influence the course of development, the majority of colour scanners worldwide have been sold to small users. Such users rarely have a clear management/worker distinction, and the manager–owners usually have a good idea of work progress and problems and whether throughput and product mix is producing a satisfactory payback. As a result, there are no management diagnostics or management controlled setup features designed into the equipment, although a minority of larger firms, significantly those with professional managers, have requested them. Thus there have been none of the fights over machine control which have taken place with the introduction of many machine tools (Wilkinson, 1983). Fascinatingly, the power relations and the relative importance of the owner/producer compared to the professional manager are reflected in producer strategy and machine design, in this case in quite the opposite way to the Bravermanian viewpoint.

CHANGES IN PRODUCTION ENVIRONMENTS

If the market relations of the colour scanner users are examined, there emerge two relatively unexplored but pervasive features of the relationship between economic growth and technical change. They are

    (i) the mutually facilitating growth of supply and demand;
    (ii) the generation of new demands based upon technical changes.

To illustrate the first point: the massive expansion of demand for colour on printed items over the past twenty years was dependent upon a group of technologies capable of supplying it, including the colour scanner. The key feature of this expansion is that the demand for scanned colour sets, the basic product, was fuelled by a constant decline in their real cost over time – a consequence of equipment costs staying more or less constant in real terms and major increases in productivity and output being obtained. Thus, virtuous circles of mutually facilitating supply and demand led both to an expansion of the market and an increase in adopters. This is a genuine bandwagon effect in the Schumpeterian sense of the kind that has been described to occur on a broader scale at the beginning of long-wave upswings (see chapter 2).

The second point can conveniently be illustrated by the changes that have happened in the early 1980s. The increase in demand has no longer been rapid enough for the increased supply from adopters and the profitability of scanning has been falling. As a result, adopters have searched for ways of maintaining viability. One of the ways of doing so is to generate new demands based upon new technological capabilities. All the firms adding electronic page make-up have found that its ability to create a more complex product which previously the market could not afford, nor necessarily envisage, has led to an increasing demand for more complex work which would not previously have been possible. Furthermore, using electronic page-make-up to satisfy the demand that has been created is probably not profitable, so firms depend upon some method of cross-subsidization or pay-back arrangement to allow them to develop a market which they expect will be profitable.

The significance of these points for the interpretation of diffusion is that the 'changing pattern of effects' of diffusion that Gold (1981) described are, indeed, mutually interrelated: adoption patterns, technical changes and market effects all affect one another as well as affecting the rate or timing of diffusion. In addition, it is clear that using profitability alone as a criterion does not provide an adequate explanation of adoption decisions – expectation may be just as important. Similar interrelationships affecting the rate and timing of diffusion can be demonstrated in factor markets: for example, in the development of labour skills, although they are not discussed here.

ADOPTERS AND POTENTIAL ADOPTERS RECONSIDERED

Having examined some of the factors affecting the pattern of diffusion, it is now possible to reconsider the changes in adopters. What is seen is a shift in advantage at the establishment level between the printer and publisher, who can spread the running cost of a piece of equipment over a whole production sequence if they believe it will confer competitive advantage, and the repro house which has no advantages of cost spreading, but does usually achieve higher utilization and produces a wider range of products by virtue of specialization in a single production stage. At the beginning of the diffusion period, the scanner was limited and restricted and only viable for large (in printing terms) integrated establishments to use. With the technical developments to the scanner made in the early 1970s, the advantage began to swing towards small establishments and small enterprises taking advantage of the equipment's versatility and flexibility. This was an advantage that became compounded by market responses: advertising agencies, publishers, mail order catalogue and brochure firms, buying sets from specialist scanning firms and putting them to a variety of printers, and printers subcontracting excess work, for example. The balance of advantage is fine, however, and slight changes may shift it. For example, a larger proportion of the electronic page-make-up systems adopted have been installed by large multiplant enterprises who see the advantage of offering customers a 'complete service', particularly in the fields of international periodical publishing. This tilts the balance once again towards integration at the enterprise level. However, this is not reflected in integration at the establishment level as was formerly the case – the pattern is now one of specialization in dispersed small establishments linked together by rapid communications.

The importance of this reinterpretation is that it shows that establishment size is not wholly exogenous to the diffusion process. The changes in the machinery permitted small establishments to adopt, whereas before this was not viable. Establishment size, however, is only an indicative measure of the properties of the firm and its production environment. In this case the critical factor seems to be the degree of specialization of the establishment: again, the developments to the scanner during diffusion have permitted specialization in a stage in the sequence of the production of printed matter.

The small highly specialized establishments that developed from the mid-1970s onwards did not, of course, form spontaneously. The first one was set up in London in 1974 by a man who can only be described as an innovator: he created a new set of market relations and a division of labour between establishments that simply had not existed before. Now why the option of small specialized establishments should have been taken up is a different matter. Scanners could have been used in multiple machine shops as numerical control machine tools often are. To explain why, the particular combinations of printing production processes and the industrial or rather social, relations of production in the printing industry would have to be examined.[3] Rather than exploring these here, the point to be made is that this approach to innovation diffusion does actually raise such important questions.

This analysis allows us to question the validity of the concept of 'potential adopters' in this case. It is clear that the technical changes that occurred during the course of diffusion resulted in shifts in the groups of actual adopters. It could not be argued that the specialist enterprises represented potential adopters at the beginning of the period. In the case of 'scanning-only' enterprises, they did not even exist! Clearly, then, a population N of potential adopters is not feasible for the whole period. An alternative might be to identify groups of potential adopters for each of the five broad phases of development under which adopters are grouped in table 3.1 – thus producing a series of curves. Elegant certainly, but how can potential adopters be identified? Should all printers above a certain size be included as potential adopters for the first scanners? A very great number no doubt could have adopted. How can the two pottery firms buying colour scanners be included or the mail order house and the department store buying page-make-up systems? Are all pottery firms and later all mail order houses and department stores to be included? How can a population of potential adopters be identified when new firm formation is taking place on the basis of technical changes? These are practical problems of a considerable order. More important is the fact that the adopters who did adopt were innovators who created new products, new interestablishment relations (divisions of labour) and new market relations. To consider that their adoption is a logical step or a matter of time misses out on this vital component of diffusion. Yet this is exactly what the identification of a population of potential adopters would do.

CONCEPT OF 'POTENTIAL ADOPTERS' RECONSIDERED

It might be argued that the colour scanner is not a good example to study since so many technical changes have taken place during diffusion. The same would also have to be said of most of the significant technological developments that have taken place: CNC machine tools or computers, for example. It is possible to identify stages of separate innovation, each with its own diffusion curve, although a little thought will reveal the problems in identifying potential adopters in the light of what has been just said. Alternatively, diffusion studies could be restricted to unproblematic innovations with a clear-cut group of adopters. (Is it unfair to see the apparent obsession with brick kilns, blast furnaces and giberellic acid in diffusion studies in this light?) In either case the dynamics of the interaction between innovation and diffusion are obscured and thus, perversely, the importance of the diffusion process, as it is revealed here, is overlooked. Mathematical elegance is sacrificed, but, if this interpretation is valid, the benefit is that the study of innovation diffusion can be more thoroughly integrated into the study of technical change.

IMPLICATIONS: SPATIAL POLICY AND PRACTICE

Just as in the first section of this chapter the economic impact of innovation was shown to have its effect through the process of diffusion, it is now clear that the process of diffusion also stimulates innovation. It is possible to be more specific than this by saying that the responses diffusion triggers, in both the characteristics of adopters and the features of the production environment, affect the further course of innovation development. Development possibilities themselves turn upon the strategy that machinery producers choose to adopt.

Rather than examining levels of uptake and drawing conclusions from there, it is more illuminating to examine the ways in which process equipment (and consumer products) are used. Are they being used to fulfil the same function, to produce the same products, or are new products being produced in different ways? If the latter, what roles have changes in markets, changes in skill availability, changes in inter-enterprise/establishment links played? How do equipment producers respond? What factors constrain their strategies? Answers to these questions have been sketched out with a limited example in this chapter; this could well be done on a larger scale. It might be added that the kinds of questions that have to be asked of producers and users to elicit this information are a good deal more interesting for both interviewer and producer than the checklist questionnaire with which both sides are all too familiar.

The particular expertise of social scientists lies in examining how the changes taking place over the course of diffusion are related to the organization of production. As reported earlier, it does seem to be the case that in certain regions in certain sectors small firms are considerably more innovative – are developing new products, new market relations and new social divisions of labour (in Pahl's (1984) sense of the term). Surely the chief contribution of social science is in looking at innovation in productive relations and how that meshes with technical innovation?

Rather than sidestepping the theme of the book by talking about future work, it is worth posing the question 'What are the effects likely to be regionally or nationally of attempting to speed up the diffusion rates of new process innovations.' The proposition is one that has looked more attractive with the view derived from interpretations of long-wave theory that installing the new technological base of the next upswing will help to bring it about (Hall, 1982; Rothwell, 1982).

Taking the example of the diffusion of the colour scanner again: it may have been possible to increase the rate of diffusion of the scanner and it may be possible to increase the uptake of the new page-make-up component – various measures to subsidize capital costs or provide training schemes for operators could be introduced. What might the effect be? Given that adoption has been based upon falling input costs and the development of a new market, that for colour images, increasing the rate of uptake may lead to the further extension of the virtuous circles of demand described earlier and, hence to an extension of demand, an increase in activity and perhaps increased employment. However, it must be borne in mind that the creation of new demand, such as that for more sophisticated images, is a gradual learning process for both final consumers, print buyers and producers. This acts as a brake on the rate of increase in demand. So, it may have been or may be possible, in this case, to increase economic activity by increasing the rate of diffusion. Can this conclusion be extended into a recommendation to increase the rate of diffusion of other currently widely diffusing technologies in order to produce a similar increase in economic activity on a larger scale and hence hasten an economic upturn?

To answer this question it must first be recognized that the example of the colour scanner is relatively unusual in that market extension has taken place during a period of recession and adoption has largely involved an addition of capacity. This is probably uncommon: what evidence there is suggests that much process innovation adoption has been to reduce costs in markets where demand is static (Freeman, 1984; Harris and McArthur, 1985). This may result in labour shedding on a scale large enough to depress demand in the economy, which, in turn, creates further problems for expansion (Pasinetti, 1981). What then if attempts are made to increase the rate of adoption of current widely diffusing technologies? They may in part create new demand sufficient to offset labour productivity increases, as seems to have been the case with the colour scanner, but it seems more likely that at present such measures may result in labour shedding in many activities. What is important is the *net balance of effects* on the level of economic activity at a regional or national level. Thus, encouraging adoption may lower the rate of economic activity in a region and increase imbalances rather than having an equilibrating effect.

The distinction between employment creating and destroying effects clearly cannot be resolved by examining the innovation's characteristics; neither can it be resolved entirely by examining the characteristics of markets and the creation of new demands (itself a form of innovation); it also revolves around the organization of production. To return to the examples from Italy and France once more, small firms with a very different set of productive relations to those of the corporation have been particularly successful in producing products and intermediate goods for international markets in small series, by taking advantage of the flexibility of computer-controlled machinery. They have tapped new sources of demand for customized rather than standardized goods and, perhaps, captured existing markets by virtue of quality and cost competitiveness. The other component of innovation then is in the relations of production; this may involve conscious 'restructuring' and it may also include a shift in the locus of competitive advantage to different forms of organization, drawing upon divisions of labour which are not automatically reproducable everywhere.

Are these new forms of organization (or perhaps old forms with a new competitive advantage, (c.f. Sabel and Zeitlin, 1985) creating employment? In the Italian cases in particular, this is hard to say. Certainly in these areas nobody talks of crisis, but whether someone works mostly in the 'black' informal or mostly in the 'white' formal economy. As Gershuny and Miles (1983) have asked, how do we measure employment?

Let us conclude by returning to the steam engine and the computer.[4] On entering the Manchester Science and Industry Museum, massive steam engines attract the visitors' gaze and their imagination. Alongside is a series of small engines – often gas-driven – that were in use in the Manchester area at the same time (early nineteenth century). Why did they have limited success when the massive, centralized, wasteful steam engine with its belts, shafts and pulleys succeeded? The microcomputer is commonly thought to have only been possible with the development of the microchip, yet the small computer providing decentralized computing power has always been technically feasible but remained undeveloped for a long time. Does the answer to these puzzles lie in the need of organizations to centralize key resources to control them, be they the capitalists of the factory system or the bureaucracies which directed the development of the computer (Jamous and Gremion, 1978)? Is this now changing?

NOTES

1 A fuller discussion of the 'high-technology' debate can be found in Harris and
   McArthur (1985) where a distinction between 'newly emerging' and 'widely
   diffusing' technologies is proposed as a means of distinguishing those technologies
   that are currently playing a considerable role in economic change and those that
   are in preliminary stages of development and currently have a limited impact,
   whatever their future potential may be.
2 The information in this section is based on research carried out as part of the
   author's PHD thesis.
3 A fuller account of these is provided in the author's doctoral thesis.
4 I am grateful to Jean Saglio for these examples.

REFERENCES

Alderman, N., Goddard, J., Thwaites, A. and Nash, P. (1982) Regional and urban
   perspectives on industrial innovation: applications of Logit and cluster analysis
   to industrial survey data. *Discussion Paper 42* Centre for Urban and Regional
   Development Studies, Newcastle.
Bagnasco, A. (1985) La costruzione sociale del mercato: strategie di impresa e
   esperimenti di scala in Italia, *Stato e Mercato* 13, 10–49.
Bagnasco, A. and Triglia, C. (1985) *Societa e politica nelle aree di piccola impresa:
   il caso di Bassano*, Arenale, Venezia.
Berg, M., Hudson, P. and Sonenscher, M. (Eds.) (1983) *Manfacture in Town and
   Country before the Factory*, CUP, Cambridge.
Braverman, H. (1974) *Labour and Monopoly Capital*, MRP, New York.
Brusco, S. (1986) Small firms and industrial districts: the experience of Italy. In
   Keeble, D. and Wever, E. (Eds.) *New Firms and Regional Development in Europe*,
   Croom Helm, London.
Courliet, C. and Judet, P. (1985) 'Nouveaux espaces de production en France et
   Italie' *Annales de la Recherche Urbaine* 29, 95–103.
David, P. A. (1975) *Technical Choice, Innovation and Economic Growth*, CUP,
   Cambridge.
Davies, S. (1979) *The Diffusion of Process Innovations*, CUP, Cambridge.
Devine, W. W. (1983) From shafts to wires: historical perspectives on electrification,
   Journal of Economic History 43, 347–372.
Dosi, G. (1981) Institutions and markets in high technology industries: an assessment
   of government intervention in European microelectronics. In Carter, C. F. (ed.)
   Industrial Policies and Innovation, Heinemann, London.
Fogel, R. W. and Engerman, S. L. (Eds.) (1971) *The Reinterpretation of American
   Economic History*, Harper and Row, New York.
Freeman, C. (1982) *The Economics of Industrial Innovation*, 2nd ed., Frances Pinter,
   London.

Freeman, C. (1984) Keynes or Kondratiev? How can we get back to full employment? In Marstrand, P. (Ed.) *New Technology and the Future of Work and Skills*, Frances Pinter, London, 13–32.

Freeman, C., Clarke, J. and Soete, L. (1982) *Unemployment and Technological Innovation*, Frances Pinter, London.

Fua, G. (1985) Les voies diverses du developpement en Europe, *Annales Economie Societe Civilisation* 40, 579–603.

Gershuny, J. and Miles, I. (1983) *The New Service Economy*, Frances Pinter, London.

Gibbs, D. and Edwards, A. (1985) Diffusion of new production innovations in British Industry. In Thwaites, A. and Oakey, R. (Eds.) *The Regional Economic Impact of Technological Change*, Frances Pinter, London, 132–163.

Gold, B. (1981) Technological diffusion in industry: research needs and short-comings, *Journal of Industrial Economics* 29, 247–69.

Hall, P. (1982) The Geography of the fifth Kondratieff Cycle, *New Society* 56: 960, 535–537.

Harris, F. and McArthur, R. (1985) The issue of high technology: an alternative view, *Working Paper No. 16*, NWIRU Manchester.

Hippel, E. A. von (1978) Users as innovators, *Technology Review* 11, 31–39.

Howells, J. R. L. (1984) The location of research and development: some observations and evidence from Britain, *Regional Studies* 18, 13–29.

Jamous, H. and Gremion, P. (1978) *L'Ordinateur au pouvoir: essai sur les projets de rationalisation du gouvernement des hommes*, Editions du Seuil, Paris.

Kendrick, J. (1973) *Postwar Productivity Trends in the United States, 1948–1969*, Princeton University Press for the NBER.

Langridge, R. (1984) Defining high technology for locational analysis, *Discussion Paper in Urban and Regional Economics* series C, No. 22 Department of Economics, University of Reading.

McQuaid, R. (1984) *Definition of high-technology industries* M4 Working Note 2.2, Department of Geography, University of Reading.

Mansfield, E. (1961) Technical innovation and the rate of imitation, *Econometrica* 29, 741–66.

Mensch, G. (1975) *The Technological Stalemate*, Ballinger, New York.

Metcalfe, J. S. (1981) Impulse and diffusion in the study of technical change, *Futures* 13, 332–38.

Nabseth, L. and Ray, G. (1974) *The Diffusion of New Industrial Processes*, CUP, Cambridge.

Nelson, R. and Winter, S. (1975) Growth theory from an evolutionary perspective: the differential productivity growth puzzle, *American Economic Review* 65, 338–344.

(1977) In search of a useful theory of innovation, *Research Policy* 6, 36–76.

O'Brien, P. and Keyder, C. (1978) *Economic Growth in Britain and France, 1780–1914: two paths to the 20th century*, Allen & Unwin, London.

Oakey, R., Thwaites, A. and Nash, P. A. (1980) The regional distribution of innovative manufacturing establishments in Britain, *Regional Studies* 14, 235–253.

Pahl, R. E. (1984) *Divisions of Labour*, Blackwell, Oxford.

Pasinetti, L. L. (1981) *Structural Change and Economic Growth: a Theoretical Essay on the Dynamics of the Wealth of Nations*, CUP, Cambridge.

Pavitt, K. (1984) Sectoral patterns of technical change: towards a taxonomy and a theory, *Research Policy* 13, 343–374.

Perez, C. (1983) Structural changes and the assimilation of new technologies in the economic and social systems: a contribution to the current debate on Kondratiev cycles, *Futures* 15, 357–75.

Rosenberg, N. (1976) *Perspectives on Technology*, CUP, Cambridge.

Rothwell, R. (1982) The role of technology in industrial change: implications for regional policy, *Regional Studies* 16, 361–370.

Sabel, C. and Zeitlin, J. (1985) Historical alternatives to mass production: politics, markets and technology in 19th century industrialisation, *Past and Present* 108, 133–176.

Saglio, J., Garroustre, P., Raveyre, M. F. and Richoilley, G. (1984) *Les systemes industriels localises*, GLYSI, CNRS, Lyon.

Scherer, F. (1982) Inter industry technology flows in the U.S. *Research Policy* 11, 227–245.

Stoneman, P. (1976) *Technological Diffusion and the Computer Revolution: the U.K. Experience*, CUP, Cambridge.

Stoneman, P. (1983) *The Economic Analysis of Technological Change*, CUP, Cambridge.

Tann, J. (1981) Fuel saving in the process industries during the Industrial Revolution: a study in technological diffusion, *Journal of Business History* 15, 149–59.

Tunzelmann, N. von (1978) *Steam Power and British Industrialization to 1860*, Clarendon Press, Oxford.

Wilkinson, B. (1983) *The Shopfloor Politics of Technological Change*, Heinemann, London.

4

Producer services and economic change: some Canadian evidence

P. A. Wood

One aspect of the 'information revolution' described by Freeman in chapter 2 has been the growth of the *information processing* sector of the economy. This in turn has contributed to a general trend towards service employment in the United Kingdom during the 1970s, a trend which has led to a heightened interest in the economic basis for this growth and the prospects for its continuation (Gershuny, 1978;) Channon, 1978; Marquand, 1979, 1983; Robertson et al, 1982; Gershuny and Miles, 1983; Daniels, 1983, 1985; Damesick, 1986; Wood, 1984; Marshall, Damesick and Wood, 1985). With 62 per cent of jobs in the service sector in 1981, it has become increasingly difficult to dismiss them as merely'non-basic','consumer-orientated', 'low productivity' functions. Some of the growth has, of course, resulted from the effects of rising consumer expenditure and welfare expectations on the demand for relatively labour-intensive personal services. An appreciable proportion, however, is in private sector 'producer services', primarily serving other production and business activities. Further, the view is becoming more widely accepted that the importance of these activities lies not simply in the employment they offer, but also in the role they perform in facilitating production, as technological change imposes a more complex division of labour both within manufacturing and between manufacturing and the services to which it is linked. Thus, while the numbers employed directly in production are falling, the numbers employed away from the factory floor are rising. The latter are engaged in the planning and design of new products and processes, the establishment and sustaining of production itself and the subsequent distribution, maintenance and servicing of goods. Such support workers may be engaged by manufacturing firms themselves or by separate 'service' organizations. Whichever is the case, their contribution to the production process may effectively be identical (Wood, 1986).

Traditional production-supporting roles, for example in distribution and maintenance, have been supplemented in recent decades by the rapid expansion of information-processing activities serving national and international commerce, trade and government as well as production. This sector has increasingly acquired its own dynamism, expanding and diversifying the specialist technical, financial, legal and business expertise that it has to offer. Nevertheless, while national economic success no longer depends on a high level of manufacturing employment it still requires a highly productive core of manufacturing output. This increasingly depends upon the effectiveness with which it is supported by and employs 'service' expertise. The information services, together with goods-handling functions such as distribution, storage, repair, maintenance and construction, therefore offer an indispensable context for modern production. At both national and regional levels, the variety and quality of the 'information environment', and of the 'servicing' and 'maintenance' environments, seem to be of growing significance to the success of productive investment (Marshall, 1983, 1985).

THE CANADIAN CONTEXT

While the relationships between producer services and processing have recently attracted increasing interest in Britain because of the apparent collapse of manufacturing employment, such relationships have been of longer-term significance to Canadian economic development. Indeed, the evolution of the Canadian economy has been supported by two pillars; primary resource extraction and the service infrastructure required to deliver these resources to home and overseas markets. The latter has acquired added significance because of the unique servicing difficulties in Canada associated with long distances and harsh environments.

In spite of this, service activities and employment have not always attracted the attention that their importance deserves (for an exception, see Rangachang, 1982). Some commentators have been primarily concerned with the low level of Canadian manufacturing activity and its apparently poor technical and organizational status. For these, a high dependence on services has tended to be regarded as a corollary of underindustrialization or, more recently, of supposed deindustrialization, and therefore a symptom of wider economic ills (Britton and Gilmour, 1978). Even where strong producer services in Canada have been recognized as an economic necessity, this has been viewed as a threat to Canadian economic independence. Wallace Clement, for example, writes of the unequal alliance between the leading elements of capitalism in Canada and the United States:

US capitalists control the major areas associated with production, the cornerstone of an industrial society; Canadian capitalists have their strength in areas of circulation and service to production...(consisting of) financial activities, transportation and utilities (Clement, 1977, p. 6).

This division is the outcome of what Clement calls 'the Canadian quandary'; both developed and underdeveloped, a resource hinterland and an advanced manufacturer, capital rich and capital poor. In the continuing debate between 'continentalist' and 'nationalist' attitudes to economic policy in Canada (Daly, 1979; Economic Council of Canada, 1983), the historically necessary role of producer services seems seldom to have been explored in any detail, whether in facilitating the development and efficiency of extractive and manufacturing activities or themselves offering a comparative advantage in trade.

The need for scrutiny of service trends in Canada is made more urgent, of course, in view of the longstanding and well documented shift to services in the rest of the continental economy (Stanback, et al, 1981; Noyelle and Stanback, 1984; Clark, 1984). US employment in all types of services is being affected by corporate and technological changes that are likely to be of greater magnitude in the future than any further changes in manufacturing. So far, however, as in Britain, producer service studies in Canada have tended to concentrate on the pattern of high-level office developments, (Semple, 1977; Semple and Green, 1983), rather than on the more fundamental and broadly based role of all types of producer service in Canadian development.

Neglect of the role and detailed structure of service sector change seems therefore to be a common feature of regional and national economic analysis in Canada and the UK. The evolution, size, structure and recent performances of the two economies do not, however, make them obvious candidates for further close comparison, and this is not intended here. Nevertheless, some of the questions raised about the role of producer services in the UK can at least partly be answered with respect to Canada, mainly because of the superior availability of information. The parallel between experience in the two countries is thus worth drawing. For example, it has already been noted that the efficient development of producer services in Canada has not only assisted indigenous enterprise. It has also made easier the penetration of goods, services and capital from outside. As the dependence of the British economy upon producer services grows, such ambiguity in its impact is highly significant, but has been little explored at either national or regional scales. Canadian evidence, no doubt like that from other countries better served by economic monitoring, may therefore indicate significant aspects of service development that require further attention in both countries. Before exploring

Canadian regional and national data on the relationships between services and production, recent economic trends are first outlined, to provide a context both for the Canadian evidence itself and for any general conclusions about their significance for the UK.

CANADIAN AND BRITISH SERVICE EMPLOYMENT TRENDS

The comparative significance of service employment in Canada and Britain and recent trends are summarized in table 4.1. In Britain, the almost precipitous decline of labour inputs to manufacturing and the apparent buoyancy of some producer service activities demonstrates clearly enough why there has been growing interest in the latter. In Canada, on the other hand, service trends must be interpreted in a different context. First, expansion of manufacturing employment continued throughout the 1970s (i.e. before the recession after 1981), even though its share of the total fell because of faster growth in services. Whatever shifts occurred in the inter-dependence of manufacturing and service employment during this period were, therefore, subsumed within the growth of both sectors. A second element in the Canadian scene is the established significance of service support to primary and secondary production. This was reflected in the high and still growing proportion of workers in distribution and the importance of transportation and communication and construction services. Unlike in Britain, these materials handling services in Canada continued to expand throughout the 1970s. By 1981, the patterns of information processing service employment had become more similar in the two countries than ten years earlier, but Canadian growth in the 1970s was more diverse than in Britain, supported by a strong expansion of public and consumer, as well as producer demand.

Another aspect of the employment structures of the two countries that is significant in assessing the scale of 'service' activities, is the occupational pattern within manufacturing and services. Table 4.2 shows this for 1981. Although the distinction can be made between manual and white-collar workers within manufacturing/construction, it is unfortunately not possible to explore the producer/consumer distinction within services. Table 4.2 nevertheless confirms the main structural differences between the two economies. In particular, the relative importance of services in Canada is seen to be concentrated into manual and clerical/sales jobs for men, rather than into professional and managerial occupations. This no doubt partly reflects the relative importance of materials handling services, noted in table 4.1. It also shows that the service sector has historically offered no

Table 4.1 Employment structure, Canada and Great Britain, 1971 and 1981

% of total	Canada 1971*	1981	% change	G.B. 1971	1981	% change
Primary	9.1	7.0	+ 6.8	3.4	3.4	–
Manufacturing	21.1	19.1	+24.5	38.1	28.0	– 29.2
Services: Information handling
Professional	16.6	18.4	+53.3	12.3	17.0	+33.4
Fire**	4.6	5.5	+65.9	3.3	6.2	+82.6
Misc. Services	9.1	10.8	+62.0	9.9	12.0	+16.9
Public Admin.	8.2	7.7	+30.0	6.4	7.1	+ 7.7
	38.5	42.2	+51.9	31.9	42.3	+28.1
Services: Materials handling
Construction	6.4	6.2	+33.0	5.7	5.2	– 12.7
Utilities	1.1	1.2	+41.8	1.7	1.6	– 8.4
Transport & Distribution	7.3	7.1	+34.0	7.1	6.7	– 9.3
Distribution	15.8	16.9	+47.2	12.0	12.9	+ 2.9
	30.6	31.4	+36.3	26.5	26.4	– 4.4
TOTAL			+37.6			– 3.6

* Includes proportional allocation to each sector of 'unallocated' workers (7.7% of total)
** Finance, Insurance and Real Estate
Source: Canadian Censuses, 1971, 1981, UK Department of Employment Gazette.

clear compensation for the deficiencies in high-level employment of Canadian primary and manufacturing activities. The proportion of manual jobs for men in the two countries is identical, but, as would be expected, more of them in Canada are in the primary and service sectors. For women in Canada, compared with G.B., clerical/sales jobs predominate over manual employment, although there seem also to be more opportunities in professional/ managerial occupations.

Occupational changes in the major sectors cannot be examined for the UK. In Canada, however, between 1971 and 1981 (table 4.3), the main changes, were towards the *consumer, business and personal service* grouping. Virtually all occupations were involved, for both males and females. The rates of growth, however, were almost invariably higher for women, especially in the managerial/sales and clerical occupations. In other sectors, expansion

Table 4.2 Occupation/industry patterns, G.B. and Canada, 1981, showing percentage of total male and female workforces

OCCUPATIONS		Primary		Manuf/Constr.		Services		Total	
		M	F	M	F	M	F	M	F
Prof/Managerial	G.B.	2	—	9	2	19	21	30	23
	Canada	1	—	4	1	18	24	23	25
Clerical/Sales	G.B.	—	—	3	8	8	32	11	40
	Canada	—	1	3	5	13	40	16	46
Manual	G.B.	4	1	33	12	21	23	58	36
	Canada	7	2	25	8	26	18	58	28
Total	G.B.	6	1	45	22	48	76		
	Canada	8	3	32	14	57	82		

(INDUSTRIES)

Source: Censuses

Table 4.3 Canada, 1971–1981, major changes in occupation/industry patterns (absolute changes, in thousands)

| | OCCUPATIONS | | | | | | | | | |
| | Manag/farm/ sales | | Prof/scientific | | Clerical | | Service | | Primary/ secondary processing | |
Sectors	M	F	M	F	M	F	M	F	M	F
Primary	0	4	16	6	2	23	1	3	69*	5
Manufacturing	58	22**	30	14	−4	58	2	3	287	141*
Construction	35**	17	6	1	−1	31	0	1	152	7
Transport	34**	14	26	10	9	67*	0	5	148	21
Trade	168	191*	9	16	30**	175*	6	9	124	19
FIRE	60*	66***	5	6	2	129*	5	4	3	1
Consum/Bus/pers. services	102*	77**	203*	342*	18	237**	126	264*	69*	18
Public admin/def.	47*	30	94**	31**	1	91*	18	20	19	3

*** > 100% increase,
** > 50% increase,
Source: Censuses

was occupationally more selective. The most consistent trends were the widespread absolute growth in *processing jobs for men* in primary, manufacturing, construction, transportation and trade activities, and in *clerical jobs for women*, especially in the services. It is noticeable that, in spite of the low proportion of managerial and sales jobs in manufacturing, their growth was modest compared with the same occupations in construction and the services, and manual jobs in manufacturing.

These data confirm the employment significance in Canada of both the materials and information processing services and their continuing growth, especially of the latter. They also reveal the relative paucity of professional/ managerial jobs in the processing sectors, but suggest that these deficiencies are now being made up by growth in certain services (table 4.3). The evidence leaves other important questions unanswered, for example concerning the balance of change *within* the consumer/business/personal service category, especially between public sector employment, and private consumer and producer services. Similarly, the apparently growing concentration of managerial/sales expertise in the service sector needs to be more fully investigated to discover its 'producer' content. It is also not known how far the expansion of manual occupations in the service industries reflects an increase in producer service subcontracting from primary/manufacturing activity, compared with other changes in the public and consumer service sectors.

These issues, of course, are unlikely to be clarified as long as classifications of services continue to ignore their market characteristics, in particular the producer- or consumer-orientation of firms' activities. Several attempts have been made in the UK to assess the relative employment levels in these categories by the division of Standard Industrial Classification (SIC) employment between notional producer and consumer markets (Marquand, 1979; Robertson, et al., 1982; Wood, 1984). Table 4.4 adopts a similar approach for Canada. In some ways, this is easier than with comparable exercises in Britain, because of the detail of the Canadian SIC and the consistency of its application to the workforce data in the 1971 and 1981 Censuses. In table 4.4, the employment for some individual industrial groups has been allocated entirely to one of the two categories; for the producer services these include freight transport and storage, wholesaling, investment dealers and business services, and for the consumer services, retail trade, road passenger transport/ urban transit, and most consumer and personal services. In cases where both consumer and producer functions are included in the SIC group, employment has been divided in proportion to the shares of the major SIC groups (construction, transportation, communications, utilities, finance, insurance and real estate, community, business and personal services) going

Table 4.4 Estimated workforce share of producer and consumer services, Canada, 1971 and 1981 (employment in thousands)

		1971	1981	% Change	% of Total 1971	% of Total 1981
Processing Industries (including agriculture & fishing)	M	1792	2192	+22		
	F	490	733	+50		
	T	2280	2925	+28	33	28
Producer Services	M	1065	1586	+49		
	F	300	692	+131		
	T	1365	2278	+67	20	22
Consumer Services	M	1676	2366	+41		
	F	1554	2734	+76		
	T	3230	5100	+58	47	50
Public Administration	M	460	549	+19		
	F	157	314	+100		
	T	617	863	+40		
TOTAL (Excl. public administration)		6875	10303			

Source: Censuses

to intermediate or final demand in the 1980 input/output table. Although there are a number of well known limitations to this type of exercise, it does at least attempt to differentiate between various degrees of the 'mixed' involvement of services with producer and consumer markets.

In 1981, the estimates suggest that 22 per cent of Canadian employment was in producer services (especially in wholesaling, business services and transportation and storage), compared with 28 per cent in the processing sectors (primary and manufacturing) and 50 per cent in the consumer services (especially education, health, retailing and personal services). Significantly, in the previous decade producer services had grown fastest, by 67 per cent including a particularly rapid increase in women's jobs and in business and miscellaneous services, with slower growth in the utilities, transportation and storage and construction. Consumer services grew by 58 per cent, but employment in the processing sectors expanded by only 28 per cent. The share of producer services seemed to be higher than in

Britain (where it was around 18 per cent), balancing to some extent the shortfall in processing employment. As a result, the share of consumer service employment, at 47–50 per cent, was about the same in the two countries, allowing for the approximation of the estimates. The recent expansion of service occupations in Canada seems therefore to have been driven to a significant degree by producer-orientated business and other support services, as well as by public sector and consumer services. Some of this growth may have taken place as a result of the 'displacement' of jobs from the processing industries, but this is certainly not the whole story. Organizational and productivity trends within the service sector itself, and its changing role in the economy at large, are probably of greater significance.

There can be little doubt from this evidence that the scale of the service sector and trends within it retain their significance for the future of Canadian employment. As in the UK the impact of technological and organizational changes, influenced by international as well as national pressures, is likely to be expressed most forcibly through the service sector. The analysis of employment data, however, disguises our ignorance of the dynamics of employment change within this sector. It also can tell us nothing about the interdependence of different sectors and the role of particular types of service in sustaining or complementing changes in other activities including production and other services. In Canada, however, the prospects of progress in answering such questions are more promising than in the UK for two principal reasons. First, as we have seen, producer services have historically been of special significance to the Canadian economy. Their role may therefore be expected to be more clearly drawn than in a traditionally manufacturing-dominated economy. Second, the quality of the Canadian evidence is unmatched.

For example, several important pieces of Canadian geographical research since the early 1970s have provided information about the service base of processing activities at the local scale. These illustrate the practical reality of producer service demand at the micro level. In addition, however, Canadian input/output evidence nationally enables the macro economic *demand* for producer and consumer services to be explored in more detail and over a longer period (nearly twenty years) than in any other country. Further, recent work employing this information for the Economic Council of Canada has offered revealing insights into the contribution to national productive *efficiency* of service changes. This is an aspect of current structural change that is impossible to address from UK sources, even though, with the growth of service activities, it is of vital concern to the assessment of UK industrial performance.

Of particular interest for geographical research in this juxtaposition of micro- and macro-evidence for Canada is the relationship between local and national patterns of service provision. In the UK for example, as in Canada, different regions have acquired different shares of information-processing producer services. Although this has direct implications for the regional structure of labour demand in the service sector, it is not so clear how indirectly damaging this differential may be to the success of manufacturing in the various areas. The Canadian evidence at least suggests ways in which a regionally varied producer service structure may support different forms of industrial development. It also raises the issue at the local scale, already pointed out nationally, of the ambiguous role of service functions in regional economic development. This will be returned to in the Conclusion.

ROLE OF PRODUCER SERVICES: CANADIAN REGIONAL AND
LOCAL EVIDENCE

The dependence of manufacturing in particular areas on inputs of producer services was first investigated in the early 1970s in several Canadian studies which, unlike linkage studies in the UK at that time, emphasized the importance of service as well as materials inputs. The pioneers were Bater and Walker, in their survey of the Hamilton metal industries, later extended to other towns in mid-Ontario (Bater and Walker, 1971; Walker and Bater, 1976). Their contention was that the availability of industrial services might beneficially influence community industrial development, although they recognized in a later review that this impact still needed to be substantiated (Bater and Walker, 1977). Their evidence nevertheless demonstrated the dependence of manufacturing plants on non-goods service inputs, especially in the form of regular transportation, repair and maintenance, and banking services from the local economy, which in Hamilton constituted over 80 per cent of the number of service links (table 4.5). On the other hand, the more specialized and infrequently used information services tended to be brought in from Toronto. Like Britton (1974, 1976), in his linkage study of a sample of firms in four industries in different parts of Ontario, Bater and Walker also emphasized behavioural factors influencing the dependence of particular plants on local services, especially their size and the single or multiplant character of firms. These propositions, of course, have only more recently attracted attention in the UK literature (Marshall, 1979, 1982). Also more recently, Polese (1982) has examined the balance between local sources of business services and interregional trade in the eastern townships of Quebec, on the basis of an unusually comprehensive survey of 24 services, supplied to various types of business (manufacturing, retailing, wholesaling,

Table 4.5 Principal non-goods linkages of 138 Hamilton area metal-working firms, by regularity of contact (ranked by number of linkages)

	Daily	Weekly	Monthly	Other*	Total
Transport companies/ warehousing	244	82	82	48	456
Repair/welding/pipe-work/ elec/subcontractor	24	41	92	114	271
Banks	118	64	7	3	192
Advert/printing/photography	8	13	53	79	153
Insurance	3	5	24	105	137
Accountants	4	5	40	87	136
Legal	3	5	27	100	135
Waste disposal	8	67	24	8	107

* Quarterly, less frequently or irregular
Source: Bater and Walker, (1971, 1977)

construction and business services). He emphasized the growth of non-local service sources for nearly all types of inputs, including even those 'market oriented' services (i.e. bought in from other firms, such as construction, real estate, repair, transportation, equipment rental and manpower training), which in the past have most commonly come from the local region (column 3 in table 4.6). Although over half of the services were imported, however, especially from Montreal 150 km away, 45 per cent of these were 'hidden' in the form of 'organization-oriented' movements between the branches of large firms. These movements were particularly characteristic of office-based financial and business services (column 2 in table 4.6). The general pattern was for larger plants to rely more on intrafirm transactions, although the outside purchase of computer and employment services contradicted this rule, tending to increase with the size of establishments.

Like the recent UK work by Marshall, these surveys suggest that the *local* income and employment multiplier effects of producer service links to manufacturing are declining, both as the supply of services becomes more nationally organized and service provision is internalized within large firms. Although measurement of the scale of these relationships is inevitably difficult, Bater and Walker made a valiant attempt to do this in two surveys, of Kingston and Sault Ste Marie, Ontario, in 1975 (Inducon Consultants, 1976a, 1976b). They computed the *values* of service purchases by manufacturing

Table 4.6 Procurement of services, 408 businesses in eastern townships of Quebec, 1980

	1	2	3	4	5	6	7
Information Services							
Employment services	0.4	21	68	28	4	—	—
Manpower training	0.5	15	70	11	1	9	10
Student training	1.5	2	98	—	1	—	6
Long-term loans	5.0	38	33	36	8	1	17
Factoring	2.9	90	5	86	1	8	—
Insurance	6.6	59	23	58	7	8	4
Real estate	—	0	89	—	11	—	—
Legal services	1.7	62	28	61	2	7	2
Accounting	3.1	62	22	39	31	7	1
Management consultants	0.4	60	25	46	7	4	18
Publicity	5.3	39	26	37	9	12	22
Marketing studies	0.4	78	3	68	1	4	24
Computer services	3.9	32	3	87	4	4	3
Landscaping/architects	0.3	31	36	16	33	—	16
Engineering consultants	7.3	12	46	49	4	—	2
Technical studies	4.5	41	3	87	1	8	—
Customs brokers	1	11	43	18	8	—	32
	44.7						
'Materials-handling' services							
General contractors	11.2	5	75	14	7	4	—
Specialized contractors	2.0	5	76	9	14	—	—
Warranty servicing	2.0	74	23	62	3	9	4
Repairs	2.4	12	48	27	18	5	3
Trucking	25.8	20	50	15	28	2	5
Rail	6.6	3	88	11	—	—	1
Equipment rental	5.3	6	35	47	13	4	2
	55.3						
TOTAL	100.0	27	44	35	12	4	5

1 Percentage of total demand
2 Percentage of demand supplied from within same firm

Percentage regional distribution of service expenditures
3 Local (Eastern Townships)
4 Montreal
5 Rest of Quebec
6 Toronto
7 Rest of Canada/overseas.
Source: Polese (1982), Tables 3, 5.

plants from the local community and farther afield, and compared these with the in-house provision of the same services. In Kingston they estimated that $46 million was spent on services (table 4.7). Only one third went to information-providing services, the rest to materials-handling services. Of the value of materials services, 39 per cent was provided 'in-house' by firms themselves, 38 per cent by local suppliers and 25 per cent from outside the area. In contrast, it was estimated that around 60 per cent of information services came from outside, including from other branches of firms, only 8 per cent from the local area and 19 per cent from 'in-house' sources.

Table 4.7 Industrial services purchased by manufacturers in Kingston, 1975 financial year, adjusted values, $ millions

	Total	In-house	Local	Outside
1 'Information' services	15.44	2.95	1.24	9.57
2 Materials-handling services	29.80	11.80	11.65	7.62
3 Manufacturing sub-contracting	0.26	—	0.06	0.17

1 Financial (5.2), Insurance (3.7), Accounting and bookkeeping (2.2), Consultancy & research (1.4), Engineering/Architecture/surveying (1.3), Legal (0.5), Property management (0.5), Advertising (0.4), Office and lab. equipment rental (0.1)
2 Transportation and warehousing (10.0), General repairs (5.2), Retailers of supplies (4.5), Building maintenance (2.7), Ground maintenance (2.1), Vehicle maintenance (2.1), machine tools and equipment (1.6), Printing (0.6), Industrial launderers (0.2), Catering services (0.1)
Source: Inducon Consultants, 1976a

As in Polese's evidence, therefore, transportation and warehousing, repair and maintenance, and wholesaling emerged as the main areas of expenditure, each divided between in-house, local and outside sources, but with a considerable emphasis on local sources compared with information-processing services. In Sault Ste Marie, the data for the white-collar functions were less comprehensive because of the domination of a single large plant. In this case, over half of the materials-handling services, however, were bought in from outside the area, and only a small proportion of all services was supplied 'in-house'. This demonstrated the local variability of manufacturing-service links, in relation to both the ownership of dominant plants and the capacities of local economies to provide services. It nevertheless also reinforces the impression from the other studies of the importance of materials-handling services in creating local multipliers from manufacturing investment and perhaps also in attracting and sustaining such investment. It confirms the significance of external ownership in reducing the local impact of information-processing services. Thus, although the attraction of office-based jobs may be a useful component of local economic development, the encouragement of the materials-handling producer services may offer rather more prospect of success, both in providing local jobs and in serving other local businesses.

Bater and Walker indicate that the local evidence for the patterns of manufacturing-service relationships raises more questions than it answers (Bater and Walker, 1977, p. 22). For example, is the lack of local information-processing capacity disadvantageous when efficient services can easily be acquired from outside, whether 'in-house' or from specialist firms? Not least problematic are those questions concerned with the impacts of technological, corporate and market changes on these patterns. Polese points to some fairly radical developments in the 'office activity complex'. The general division between in-house and outside sources of much service support to production is also changing. Thus, whatever assistance may be given to regional economic development by the local availability of materials or information-processing services, the supply patterns of such services and their role in production are not static. Any further studies of manufacturing/service relationships at these geographical scales must therefore assess the impacts of changes operating both in the national economy and in spatial patterns of corporate organisation. In Canada, fortunately, at least the first of these dynamic issues can be explored rather further.

ROLE OF PRODUCER SERVICES: CANADIAN NATIONAL EVIDENCE

In Britain, attempts to examine the output of services in relation to inter-mediate and final demand, by recourse to the national input/output tables, and thus to identify their producer and consumer roles, have not been very satisfactory. The generalized treatment of the service industries in the published UK tables and the out-of-date and static nature of the information have offered only shapshot estimates (Marquand, 1979, Robertson et. al., 1982). In 1973, for example, it was estimated that an average of 34 per cent of the value of service industry output was sold to other intermediate sectors, rather than to 'final demand' (consumers, government, capital formation, exports). This proportion was higher for communications (54 per cent) and transportation services (46 per cent). In fact, several major heavy industries spend more on transport services than on domestic inputs from any other manufacturing order. The distribution sector was also important for the food, metal goods, engineering and paper and printing industries. The propor-tion of the value of total industrial output which came from these producer services was 16 per cent compared with 29 per cent from services to consumer and government 'final demand', and 50 per cent from manufacturing (Wood, 1984).

Canadian input/output information allows a much more detailed investigation of these aggregate service/production relationships. First, it is available in the form of commodity/industry transaction tables, in which the values of purchases by major sectors of a number of service 'commodities' can be identified. Second, the data have been available annually since the early 1960s on a comparable basis, the most recent table (in 1985) being for 1980. Finally, the information is published on a standard (1971) price basis. Thus services can be identified not only as commodity inputs to various intermediate sectors and to final demand, but changes can be traced in the share of these and other inputs over almost twenty years.

Table 4.8 identifies three types of *commodity* purchases made by various economic activities in both the intermediate (A–C) and final demand (D–G) categories; I 'manual' services, mainly materials handling, utilities, construction and delivery skills, II 'information' services, including financial and business services, communications, operating services to business and government, real estate and travel activities, and III materials purchases. Enterprises purchasing in the intermediate sector (assumed for the services to constitute the main element of producer service demand) are divided between, A materials-processing enterprises, B materials-handling enterprises and C information-handling enterprises. 'Business' purchases in final demand (F) are also interpreted as part of the producer service market.

Perhaps most striking, in spite of the great simplification of the data, is the complex interdependence of these various activities, including a high dependence of the service sectors themselves on service purchases. Thus, the materials- and information-handling service sectors (B and C) purchased 23 per cent of the gross value of manual service output and 35 per cent of the information services as intermediate demand. In addition, 21 per cent of manual services (compared with only 2 per cent of the information services) were sold to other businesses (F) as 'final demand' mainly from wholesale and construction firms. In comparison, direct service purchases by extractive and manufacturing industries (A) accounted for only 9 per cent of manual service and 19 per cent of information service sales. Although these purchases accounted for 31 per cent of processing inputs, it seems that the direct role of material processing *per se* in driving the producer service economy is subsidiary to the significance of specialized intermediate services themselves, for example in finance, communications, property dealing, travel, construction, or transportation. The increasingly specialized and subdivided nature of the service sector, and the emergence of organizations able to serve both

Table 4.8 Commodity inputs of major intermediate and final demand sectors, Canada, 1980 (Percentages of sector purchases (rows) and commodity output (columns))

PURCHASING SECTORS	*I* Manual services	*II* Information services	*III* Materials	
Intermediate				
A Materials processing	12.9	18.7	68.4	/100
	9.0	19.3	43.8	
B Materials handling	36.3	27.6	36.1	/100
	13.2	14.9	12.1	
C Information handling	28.8	38.0	33.2	/100
	10.2	19.9	10.7	
Final demand				
D Consumers	37.5	27.0	34.6	/100
	34.7	38.2	29.2	
E Government	54.1	35.5	10.3	/100
	7.6	7.4	1.3	
F Business	61.9	4.5	33.5	/100
	21.1	2.3	10.5	
G Trade				
	4.0	−2.0	−7.6	
	100.0	100.0	100.0	

Manual Services: construction, transportation and storage, transportation margins, utilities, wholesaling, retailing, personal/miscellaneous services
Information Services: communications, real estate transactions, finance, insurance and real estate, business services, office/laboratory/food services, travel/advertising/promotion

producer and consumer markets, thus make the significance for production of service provision all the more difficult to examine in isolation from wider shifts in the structure of the economy. The relationship also operates in the other direction; more than one third of the gross inputs to both types of service activity was in the form of materials, for which these services clearly offer a valuable and growing market (table 4.9).

The importance of the manual services, as here defined, is reflected in their 36 per cent share of the gross value of inputs to the Canadian economy, compared with about 40 per cent from materials and only 24 per cent from information services. Within the domestic economy (i.e. excluding trade), including final sales to business, they may be regarded as 56 per cent

Table 4.9 Changes in commodity input to major intermediate and final demand sectors, Canada 1962–80 (at standard prices, 1960 = 100)

PURCHASING SECTORS	COMMODITIES			
	I *Manual* *services*	*II* *Information* *services*	*III* *Materials*	*Total* *domestic* *inputs*
Intermediate				
A Materials processing	234	244	217	223
B Materials handling	237	261	209	231
C Information handling	219	361	237	265
Final demand				
D Consumers	215	236	200	215
E Government (net)	190	365	64	184
F Business	259	256	343	282
G Trade (net)	355 (Expt)	411 (Impt)	455 (Impt)	

'producer orientated', compared with 55 per cent for the information services. In contrast, of course, materials inputs were more producer-orientated, 72 per cent in all, with 44 per cent of gross output going for further processing in the intermediate sector. Final sales of materials to consumers and government, mainly in the form of food, vehicles and household appliances, made up the remaining 28 per cent of their domestic output.

These data do not provide a full picture of the symbiotic and overlapping relationships between materials processing, producer services and consumer services; between information- and materials-handling activities or between the public and private sectors. They nevertheless indicate their interdependent nature in Canada and particularly the complexity of relationships between service functions, as well as between them and processing and consumer activities.

Over the period 1962–1980, increases in the real value of inputs to both intermediate and final demand were greatest for the information services (table 4.9), most markedly to private information-handling enterprises themselves and to government (with increases of more than 3.5 times in real terms). There was a similarly high increase in the purchase of finished materials and equipment by business, perhaps reflecting a greater reliance on specialist suppliers for more sophisticated products. The most striking trend in demand for manual services (including utilities) was in the export balance. As commented earlier, this suggests a Canadian comparative advantage in these activities, even if it may have been deployed to some degree to increase the deficits in manufactured goods and information services! In addition, however, it is noticeable how dependent was growth in manual services demand upon materials processing, other materials-handling and business-purchasing activities, rather than the faster expanding information services. The processing industries moved towards a greater direct dependence on outside services in relation to materials inputs, showing a more balanced emphasis on manual as well as information service inputs than the equivalent growth of either information-handling services or final consumer and government demand.

Again, table 4.9 no doubt disguises many complexities. Compared with patterns of final consumer demand, the main locus of growth was nevertheless clearly in intermediate and government final demand for *information services*. The main expansion of demand for the *manual services* was in trade and domestic business sales. The most buoyant markets for *materials* were in business purchasing and the information-handling industries. In this context, the growth of services sold directly to the processing sector was modest. Nevertheless over the 18 year period, for every unit of materials purchased, an extra 12 per cent of information and 8 per cent of manual services were also bought in. The impression is of an economy in which structural changes are increasingly being driven by trends in the information-processing services which, in output terms, still remained relatively small. As a result, patterns of final demand now contain significantly higher service inputs in relation to material commodities than in the early 1960s.

SERVICES AND PRODUCTIVITY CHANGES

Compared with the UK, these Canadian data allow a much clearer picture to emerge of the growing value and complexity of *demand* for services in the economy at large. A second, and even thornier problem concerns the *efficiency* of their contribution. This problem underlies any debate on the contribution of services to economic success at the regional or local scale, and careful local studies, as well as better theory, are needed to illuminate the discussion. Measurement of the productivity of service activities is notoriously difficult, partly because of the problems of evaluating both their inputs and outputs. More significantly still, much discussion of services ignores their *strategic* role in relation to production, consumption and to other service activities. The very classification of services in the SIC assumes that they peform *separate* functions from production and consumption. In fact, services are not like manufacturing industries, performing particular technical transformations of their inputs to produce discrete products. Their role is quite different; they supply expertise and enhance the value of *all other* sectors' outputs, including other services. Their economic contribution, including that to efficiency, can therefore only be evaluated in relation to the improved performance they bestow on the operations of their customers. 'Productivity', as an attribute of a sector in isolation from its wider economic role, is an even less satisfactory concept for services than for manufacturing industries. Fortunately in Canada this problem has recently been directly addressed, on the basis of the exceptional quality of the input/output data, in a study by Postner and Wesa (1984) for the Economic Council of Canada. This takes a 'total factor productivity' approach to the measurement of productivity growth between 1961 and 1978. In this, all the indirect inputs from other sectors, as well as the direct inputs needed to produce and deliver a unit of each industry's output for final consumption, are taken into account. The approach therefore seeks to identify within the economy as a whole those productivity improvements that might have the greatest effect on overall efficiency. In total productivity terms, therefore, the efficiency contribution of 'capital-intensive' industries, for example, is shown to be lower when the labour intensiveness of their intermediate inputs and capital replacement needs are allowed for. The general effect is to reduce the differences between these 'high productivity' sectors such as mining, transport equipment, chemicals, communications and utilities, and the traditionally more labour-intensive industries such as textiles, clothing, furniture, construction and, in particular, many of the services.

Postner and Wesa explore the direct and indirect effects of productivity improvements in each sector between 1961 and 1976. Table 4.10 ranks sectors from highest to lowest in terms of their 'output effects'. It is extracted from a much longer list and identifies only the top 15 in the ranking and the bottom three. The 'output effects' of productivity improvements in some sectors upon others were often of greater general significance than their direct impacts (column 5). Total productivity changes in the construction, paper, primary metals and food and beverages industries, for example, were more dependent on the input effects from other sectors (column 3) than on their own internal productivity improvements (column 2). Between 1961 and 1976, the main contribution to total productivity improvements in Canada were made by the agricultural and transportation and storage sectors, together with important contributions from forestry, construction and a number of other processing and service sectors ranked in column 4. Conversely, the main slowing effect on total productivity improvement came from the business services and other personal service sectors.

Postner and Wesa pursue the analysis by undertaking several exploratory

Table 4.10 Changes in total labour requirements attributed to 'own effects' and 'input effects', and transmitted to other industries as a result of 'output effects' (1961–76, man-years per 1 million dollars of output, ranked in order of output effects)

	1 Total change	2 Own effect	3 Input effect	4 Output effect	5 **
Agriculture	−136	−109	−27	−264	71
* Transport & storage	−58	−39	−18	−108	73
Forestry	−41	−29	−13	−56	66
* Construction	−33	−11	−22	−50	82
Paper and allied	−29	−9	−20	−36	79
Textiles	−76	−43	−33	−35	45
* Wholesale trade	−40	−25	−15	−34	58
Primary metals	−26	−6	−20	−30	84
Metal mines	−13	−7	−6	−30	81
Wood products	−49	−23	−25	−28	55
* Communications	−71	−47	−25	−25	35
Food and beverages	77	−18	−59	−24	−57
* Retail trade	−49	−31	−18	−22	41
Electrical products	−53	−29	−23	−22	42
Printing & publishing	−33	−21	−12	−22	51
* Business services	−8	−4	−5	+ 23	120
* Other personal sers.	+ 40	+ 49	−9	+ 15	23
Transportation equip.	−56	−32	−23	+ 6	(−20)

* Output effect as percentage of output + own effects
** Services
Own effect: change in total labour requirements (unit of output of a sector arising from technical change in that sector. Changes include those affecting direct labour inputs and intermediate inputs and capital replacement needs, expressed in terms of their direct labour content.
Input effect: changes in labour, intermediate and capital replacement inputs in *other* sectors, as they affect the total labour requirements of a purchasing sector.
Output effect: effect on total labour requirements in other sectors of productivity changes in a supplying sector.
Total productivity change (labour requirement)/unit of delivered output = Own effect + Input effect.
Source: Postner and Wesa (1984), tables 3.1 and 3.2.

studies of the possible impacts of various productivity improvements, significantly emphasizing the services. In one of these, an improved measure of output growth in the *finance, insurance and real estate* sector is employed, mainly to account for high inflation in the 1971–6 period. As the most important supplier of intermediate commodities, it is argued that the measurement of productivity in this sector is especially critical. The postulation of a productivity improvement of 3.3 per cent, instead of the 0.7 per cent loss suggested in earlier studies, results in a widespread effect on total factor productivity, increasing Canadian productivity growth by one quarter, from 2.0 to 2.5 per cent (p. 25).

A further exercise postulated a 25 per cent productivity improvement from 1976 to 1980 in *five service groups*; transportation and storage, communications, wholesaling, finance, insurance and real estate, and business services. No less than 60 per cent of the resulting total productivity change would have been felt in other sectors, especially in mineral fuels, petroleum, coal products, trade, primary metals, transport equipment, forestry, paper, and chemicals, and through indirect effects on the five sectors themselves. Postner and Wesa conclude that this demonstrates the potential for improvement in aggregate productivity through developments in these key services, not directly involved in serving final demand (p. 27).

In another detailed analysis of the 1973–8 period, during which productivity growth in Canada stagnated, Postner and Wesa show how an important trend at this time was a marked shift to the consumption of intermediate service inputs. There was, however, little evidence that productivity shortfalls in these services themselves affected overall productivity trends, as has sometimes been suggested. Rather, purchasers failed to complement their greater use of outside services by proportionately reducing their labour forces or inputs of other intermediate commodities (p. 44). One suggested explanation of this failing was the delay in adjusting production arrangements, following the purchase of outside service expertise, especially during a period of rapid technological change such as the introduction of new communications and computer equipment. Another suggestion is that high inflation at the time led to increased demand for outside service expertise to bring business costs under control. While emphasizing the growth and significance for modern productive efficiency of 'bought-in' services, therefore, Postner and Wesa suggest that there are productivity inefficiencies, at least in the early stages, if the purchasing sectors are slow to adapt to their use. More generally, they also show the widespread significance of service sector efficiency for the Canadian economy and highlight sectors which have a special status in the study of future national productivity trends. While in the past these have been agriculture and transportation/storage, the role of communications, finance, insurance and real estate, and business services are regarded as particularly significant for future Canadian competitiveness (p. 52).

Postner and Wesa demonstrate clearly that the efficiency contribution of some services to the Canadian economy in the past has been considerable and that there is potential for other services to make a continuing contribution in the future. What geographical distribution of nationally 'efficient' services might result, of course, involves a completely different set of considerations, which require further investigation. *A priori*, however, the importance of the 'producer service' role in Canada, and its changing form since the 1960s, is strongly confirmed by Postner and Wesa's evidence.

Even at the aggregate level, however, important questions remain unanswered. One concerns the dynamics of 'self-sustaining' growth within the service sector itself. What are the limits to this growth in relation to the growth of domestic processing industries, trade or consumer or public spending? Second, how far do the 'supply side' effects of efficient services actually facilitate import penetration, compared with the assistance they provide to the competitiveness of domestic goods and services? Better insights into these processes at the national level are probably a prerequisite for pursuing the same questions at the regional scale.

CONCLUSION

The Canadian evidence presented here has allowed the national significance of producer services to be explored more fully than is at present possible in the UK. At the local scale, Canadian research has also gone somewhat further in identifying the dependence of production upon different service inputs. Several conclusions are undoubtedly applicable on both sides of the Atlantic. First, the evidence for the dependence of manufacturing on local producer services portrays a mixed picture. The concentration of the supply of these services into certain core regions is becoming more marked, but the materials-handling services, upon which more is generally spent as part of manufacturing operations, are less susceptible to this trend than is information processing. The local economic significance of the materials-handling services, therefore, such as transportation, distribution, storage, repair and maintenance, seems to deserve more attention than it has received in the past, compared to the perhaps more glamorous 'white-collar' activities.

A growing regional inequality in the provision of information service jobs seems most likely. The pattern of provision of materials-handling services, on the other hand, is more closely related to the success of production activities in different areas. This may in turn be significantly affected by the quality of such 'blue-collar' services. Both will however be increasingly influenced by the impacts on local manufacturing of the concentration of high-quality information services into certain regions. These impacts need not necessarily all be negative, given the prospects for the efficient long-distance communication of information. The dependence of manufacturing on purely local sources of information is certainly unlikely to be beneficial. Plants owned by large firms with efficient internal channels of communication may however have a growing advantage over smaller, locally based establishments. Again, in terms of local economic promotion policies, the quality of 'blue-collar' activities may be more susceptible to planned development and sustain more local employment.

The analysis confirms for Canada the apparently 'autonomous' growth of the information-processing sector, and its significance in creating the wider environment of productive activity. This growth, as we have seen, is not directly based on producer-related trends. It depends more generally on the enhanced value of specialist expertise, the growing division of activities offering specialized service products in both producer and consumer markets and, in employment terms, the resistance of these activities to job displacement through technical innovation. Nevertheless, employment growth in the services is neither universal nor inevitable. Some types of service provision have already suffered major job losses in the UK, and they are also likely to do so in Canada. One feature of Canadian employment in the information services appears to be the high proportion of 'middle range' clerical/sales jobs, especially for women. These might be expected to be both less orientated to dominant city locations than professional/managerial occupations, and more vulnerable to technological displacement in the coming years. Corporate consolidation may speed these processes in some directions.

Perhaps the most significant evidence presented here is to be found in Postner and Wesa's work. For perhaps the first time, the essential contribution of service activities to production has been demonstrated and its consequences for national productive efficiency followed through. It has been argued here that the economic significance of any service function can only properly be evaluated in relation to the contribution it makes to other activities. The 'total factor productivity' approach made possible by the quality of Canadian input/output data allows this to be done, at least in broad terms. Of course, some aspects of the role of services are not addressed by this form of analysis. These include, for example, the part played by 'internalized' services within large companies. The method also begs many familiar questions about the adequacy of conventional measures in reflecting the benefits derived from services as well as their cost. Nevertheless, the study very effectively challenges the conventional view taken of services, in which they are still often regarded as distinct from materials-processing functions and, at best, as of only subsidiary importance for economic growth and efficiency. Postner and Wesa demonstrate the economic significance of changes in the service sector for national productivity in the past and their likely growing influence in the future.

Unfortunately there is little prospect of exploring such issues from available data at the regional or local scale. The part played by different types of service in facilitating other local economic activities probably varies widely in relation to the organizational and technical qualities of such activities and the service capacities of local economies. Postner and Wesa's analysis, however, directs attention towards the need to assess the contribution of both materials-handling and information-processing services to the competitiveness of other basic activities in different areas. This seems to be the most worthwhile direction in which to develop the tradition of Canadian geographical research summarized earlier in this chapter. In the UK this type of analysis is not technically possible at present, even at the national scale. Local and subregional studies of the role of producer services can, however, still benefit from the insights offered by the Canadian evidence outlined here.

In the future it is probable that even more rapid technological and organizational changes will take place in service provision than in manufacturing. Through their employment impact and influence on how processing activities are carried out, these changes are likely to have major effects on the form, location and social impact of processing investment. This situation does not alter the focal economic importance of materials-processing activities in economic development, both through the worth of their output and their significance for the development of technology. Nor in suggesting the existence of autonomous trends in the service sector should the fundamental need to relate service output to its ultimate producer and consumer value be neglected. On the contrary, the understanding of how service activities combine with others to serve diverse markets, and the dangers of divorcing the study of many services from their productive role, have been central themes of this chapter.

Over the past two decades, one of the main innovations in the understanding of locational dynamics has been recognition of the national and local economic significance of patterns of corporate ownership and change. Equally important now is a growing realization of the central role of service activities, not only for the employment they provide but also for their contribution to the effectiveness of materials-processing activities. The manufacturing legacy of Britain had slowed the acceptance of this fact of life, although the light now appears to be dawning. In Canada the service sector has for long been economically more significant, and clearer evidence is available for its contemporary role. It therefore offers a fruitful environment for research on the implications of the changing character of production for the complex of functions that supports it.

REFERENCES

Bater, J. H. and Walker, D. F. (1971) *The Linkage Study of Hamilton Metal Industries*, Hamilton Planning Department (see account in Walker, D. F., 1977).

Bater, J. H. and Walker, D. F. (1977) Industrial services: literature and research prospects In Walker, D. F. (Ed.), *Industrial Services*, 1–24.

Britton, J. N. H. (1974) Environmental adaptations of industrial plants: service linkages, locational environment and organization. In Hamilton, F. E. I. (Ed.), *Spatial Perspectives on Industrial Organization and Decision-making*, Wiley, London, 393–90.

Britton, J. N. H. (1976) The influence of corporate organization and ownership on the linkages of industrial plants: a Canadian enquiry, *Economic Geography* 52, 311–24.

Britton, J. N. H. and Gilmour, J. M. (1978) *The Weakest Link – A Technological Perspective on Canadian Industrial Development*, Science Council of Canada, Background Study 43.

Channon, D. F. (1978) *The Service Industries: Strategy, Structure and Financial Performance*, Macmillan, London.

Clark, G. L. (1984) The changing composition of regional employment, *Economic Geography* 60, 175–93.

Clement, W. (1977) *Continental Corporate Power: Economic Linkages between Canada and the United States*, McClelland and Stewart, Toronto.

Daly, D. J. (1979) Weak links in "The Weakest Link", *Canadian Public Policy* 5, 307–17.

Damesick, P. D. (1986) Service industries, employment and regional development in Britain: a review of recent trends and issues, *Transactions, Institute of British Geographers New Series* 11, 212–226.

Daniels, P. W. (1983) Service industries: supporting role or centre state? *Area* 15, 301–10.

Daniels, P. W. (1985) *Service industries: A Geographical Appraisal*, Methuen, London.

Economic Council of Canada (1983) *The Bottom Line: Technology, Trade and Income Growth*, Ottawa.

Gershuny, J. (1978) *After Industrial Society? The Emerging Self-service Economy*, Macmillan, London.

Gershuny, J. (1983) *Social Innovation and the Division of Labour*, Oxford University Press, Oxford.

Gershuny, J. and Miles, I. D. (1983) *The New Service Economy*, Frances Pinter.

Inducon Consultants, Bater, J. H. and Walker, D. F. (1976a) *The Role of Service Industries in Industrial Development in Kingston*

Inducon Consultants, Bater, J. H. and Walker, D. F. (1976b) *The Role of Service Industries in Industrial Development in Sault Ste. Marie*. Reports to Ontario Ministry of Industry and Tourism (see account in Walker, 1977).

Marshall, J. N. (1979) Ownership, organization and industrial linkage; a case study in the Northern Region of England, *Regional Studies* 13, 531–7.

Marshall, J. N. (1982) Linkages between manufacturing industry and business services; *Environment and Planning A*. 14, 1523–40.

Marshall, J. N. (1983) Business service activities in British provincial conurbations, *Environment and Planning A*. 15, 1343–60.

Marshall, J. N. (1985) Business services, the regions and regional policy, *Regional Studies*, 19.

Marshall, J. N., Damesick, P. and Wood, P. A. (1985) Understanding the location and role of producer services, Paper to Regional Science Association, British Branch, Manchester, September.

Marquand, J. (1979) *The Service Sector and Regional Policy in the U.K.*, Centre for Environmental Studies, Research Series, No. 29.

Marquand, J. (1983) The changing distribution of service employment. In Goddard, J. B. and Champion, A. G. (Eds.), *The Urban and Regional Transformation of Britain*, Methuen, London, 99–134.

Noyelle, T. J. and Stanback, T. M. (1984) *The Economic Transformation of American Cities*, Rowman and Allanhead, Totawa, N.J.

Polese, M. (1982) Regional demand for business services and interregional service flows in a small Canadian region, *Papers, Regional Science Association* 50, 151–63.

Postner, H. H. and Wesa, L. (1984) *Canadian Productivity Growth: an Alternative (Input-output) Analysis*, Economic Council of Canada, Ottawa.

Rangachang, U. K. (1982) *The Growth of the Service Sector in the Canadian Economy*, Ministry of State for Science and Technology, Ottawa.

Robertson, J. A. S., Briggs, J. M. and Goodchild, A. (1982) *Structure and Employment Prospects of the Service Industries* Department of Employment, Research Paper No. 30, London.

Semple, R. K. (1977) The spatial concentration of domestic and foreign multinational corporate headquarters in Canada, *Cahiers de Geographie du Quebec* 22, 33–51.

Semple, R. K. and Green, M. B. (1983) Interurban corporate headquarters relocation in Canada, *Cahiers de Geographie du Quebec* 27, 72, 389–406.

Stanback, J. M., Bearse, P. J. Noyelle, T. J. and Karasek R. M. (1981) *The New Service Economy*, Allanfield, Osmun, Totawa, N.J.

Walker, D. F. (Ed.) (1977) *Industrial Services*, University of Waterloo, Department of Geography, Publication Series, No. 8.

Walker, D. F. and Bater, J. H. (1976) *A Study of the Linkages of Metal-working Plants in Midwestern Ontario*. Report to Regional Development Branch, Ontario Department of Treasury and Economics.

Wood, P. A. (1984) The regional significance of manufacturing-service sector links: some thoughts on the revival of London's Docklands. In Barr, B. M. and Waters, N. M., *Regional Diversification and Structural Change*. B. C. Geographical Series, Tantalus Research, Vancouver, 168–84.

Wood, P. A. (1986) The anatomy of job loss and job creation: some speculations on the role of the 'producer service' sector, *Regional Studies* 20, 37–46.

5

Technology and occupational change in the United Kingdom shipbuilding industry

R. T. Harrison

The impact of new technology such as that associated with the 'Information Revolution', upon the structure of employment within national economies, is paralleled by changes at sector, enterprise and establishment levels as traditional skills are destroyed and new functions created. This chapter explores the link between technology and occupational change in the UK shipbuilding industry in the 1960s and 1970s. This is an appropriate choice. As an earlier study of the important transitions from wood to iron and steel and from sail to steam in the shipbuilding industry in the late nineteenth century made clear, the diffusion of new technology and its employment impact can be a slow and haphazard process (Harrison, 1983). More generally, the dramatic growth of new shipbuilding capacity in Japan and the Third World has destroyed the myth that long apprenticeships and inherited skills are essential for the development and maintenance of a significant shipbuilding industry (Venus, 1972; Albu, 1976; 1980). The UK shipbuilding industry, however, remains heavily craft dominated in terms of both the structure of the labour force and the structure of the trade unions represented in the industry (Hogwood, 1979, pp. 54–56; Brown and Brennan, 1970a; 1970b); changes in the relative ranking of occupations have often led to problems for the division of labour in specific localities which compounds the problems caused by the wider change in the composition and range of occupations across the economy (McGoldrick, 1984). There has also been considerable interest in recent years in the general role of technical change as a determinant of the pattern of output and employment of the shipbuilding industry at both world (Al-Timimi, 1976; Harrison, 1983) and national (Todd, 1983) levels.

CONCEPTS, DATA AND DEFINITIONS

It is a fundamental axiom of labour economics that employment changes in response to exogenous changes in aggregate demand depend, *inter alia*, on the degree of skill. Most studies of this relationship have concentrated on identifying the differential cyclical responses of various skill groups (Rosen, 1968; Nissim, 1984a: 1984b): changes in aggregate demand are effectively assumed as exogenous. In this chapter this argument is taken one step further by the examination of the effect of technical change, as one such determinant of the level and composition of aggregate demand, on occupational changes in employment.

At the outset, it is important to note that technology and technical change are complex concepts for which, despite a burgeoning literature (Henwood, 1984), there are still no comprehensive and agreed definitions either of the nature and dimensions of technical change itself or of its relationship with the level and structure of employment. For the purpose of this chapter, therefore, technical change in all its manifestations – inventions, imitation, adaptation or adoption (which together constitute innovation) – is understood in an extensive, rather than restricted sense (ACARD, 1978): innovation and technical change includes the improvement and development of both existing products and processes, and the introduction of novel production methods and products based on new technology.

Identification of the relationships between technical change in this sense and occupational changes in the structure of employment is facilitated by the adoption of a functional view of production (Hussain, 1983). The process of production, in this view, can be understood as the performance of a set of interrelated tasks or technical functions. This is not necessarily specific to a particular industrial sector or subsector: as Rosenberg (1976) has demonstrated in his discussion of the metal working industries there is considerable overlap of tasks between branches of production. However, it does follow from this perspective that a technique can be defined as a combination of tasks to be performed to achieve a particular specified result and that technical change involves an alteration in the combination of old and new tasks and the methods used to fulfil them. The importance of this view in the present context is that it facilitates the specification of a link between techniques, technical change and occupations. Specifically, technical change as defined here represents a change in the range of functions to be performed and the methods used to perform them. Furthermore, there is considerable overlap in tasks and functions between different branches of production. Consequently the process of technical change is never immediate, total or complete. It may only affect certain functions in any particular branch of production, leaving others entirely unchanged, and its impact will be influenced by the rate of diffusion of the change and by the barriers to that diffusion.

Despite this, however, these changes may have profound implications for particular occupations during the transition from the pre-change to the post-change state. By making redundant, or radically altering, old functions and creating new functions, technical change can destroy old occupations and generate new ones. But, significantly, technical change also changes the relative importance of functions and hence the relative importance of the occupations to which they have given rise.

These ideas are examined with reference to developments in the UK shipbuilding industry during the 1960s and 1970s. Unfortunately, the analysis in this chapter cannot be comprehensive because of the limited availability of detailed employment data by occupation both nationally and regionally, which is reflected in the general absence of discussion of shipbuilding labour markets in the UK. There are three readily available sources of information on the occupational structure of the shipbuilding industry in the UK. The only regularly published data relate to the two-fold Census of Production classification into operatives (all manual wage earners in the industry) and others made up of working proprietors, administrative, technical and clerical staff (ATC). This information, however, is not available on a comprehensive regional basis for confidentiality reasons, and is in any case too highly aggregated to be of value for present purposes.

However, between 1964 and 1975, shipbuilding and ship repairing and marine engineering were included in the Department of Employment survey of occupations in the engineering industry. Because of the different production technologies involved, this chapter does not include discussion of the marine engineering industry. For May of each year, a classification of employment into five broad categories was provided:

1 managerial, administrative, technical and clerical staff
2 foremen and supervisors (this category was introduced in 1973 following the revision of the occupational classification to be compatible with the revised list of key occupations for statistical purposes)
3 skilled craftsmen
4 other production occupations (predominantly semi-skilled workers, but prior to 1965 also including unskilled production workers, mainly labourers)
5 other occupations (before 1965 this category was not recorded separately from other production occupations).

The Department of Employment data also give a more disaggregated view of occupational structure in the British shipbuilding industry over this period. A total of 25 separate occupational groups within the skilled and semi-skilled craft occupations have been regrouped into two categories – shipyard trades and outfitting trades. This is to allow an examination of two hypotheses based on the results of earlier studies (Robertson, 1954; Hunter, 1967; Harrison, 1985). The first hypothesis is that employment variability will be lower in skilled than unskilled occupations. The second hypothesis is that employment variability will be lower in shipyard than in outfitting trades within the skilled craft occupations.

Although this classification of total shipbuilding employment into five categories is available for the period 1964–75, the more detailed occupational data are available only for a more limited period. First, because of a major revision to the classification of occupations in 1973, no data are available at this disaggregate level of analysis beyond 1972. Second, as already noted, semi-skilled and unskilled occupations were not classified separately by the Department until 1966. Accordingly the comparison of employment stability in the aggregate skilled and semi-skilled occupational groups can only be carried out for the period 1966–72. Third, although most occupational data for the industry relate to 1964 and after, detailed data for individual skilled occupations are available for the period 1963–72. Because of the interest in this chapter in the role of technical change on occupational change, the full range of data available has been used where possible. Finally, in addition to the Census of Production and the Department of Employment occupational survey, the decennial Census of Population includes a detailed employment by industry classification which alone enables a comparison of the occupational structure of the regional and national shipbuilding industries.

AGGREGATE CHANGES IN OCCUPATIONAL STRUCTURE

The largely craft-based nature of shipbuilding employment is reflected in the way work is administered within the shipyards, which in Britain are still characterized by a relatively small staff and an organization and a division of labour and allocation of work on the basis of occupation which gives the work group a good deal of autonomy (Brown and Brennan, 1970a; 1970b). In 1972, for example, the staff to labour ratio in the major European shipbuilding industries was 1:3 compared with 1:5 in the UK (Booz, Allen and Hamilton, 1973). This reflects differences in both staffing practices (e.g. greater use of technical staff services in European yards) and technical awareness (e.g. extensive capital investment by the European companies to reduce the relative importance of work-intensive operations in the overall cost structure).

From the detailed occupational surveys conducted by the Department of Employment it is clear that the greater importance of operatives in the shipbuilding industry reflects the considerable reliance of the industry on skilled employment (table 5.1). Even allowing for changes in the occupational classification over time, craftsmen accounted for over half of all shipbuilding employment in every year of the study period. Variations in the relative importance of the main occupational groups in the industry reflect in a very general way the pattern of technical change and its influence on labour demand (Harrison, 1985). Over the twelve year period for which comparable data are available employment in managerial and ATC occupations increased by over 20 per cent: not only was the employment share of this occupational

Table 5.1 Occupational structure of the shipbuilding and shiprepairing industry, 1964–1975 (in thousands)[a]

	Managerial, administrative technical clerical	Foreman supervisors	Craftsmen	Other production occupations[b]	Other occupations[c]	Total
1964	16.7		60.8	32.3		109.7
1965	17.2		64.9	33.6		115.7
1966	17.9		66.0	13.5	21.6	119.0
1967	19.9		67.5	13.2	21.4	122.0
1968	21.4		68.3	13.2	20.2	123.2
1969	20.9		65.9	12.7	20.1	119.6
1970	21.4		66.4	13.1	19.5	120.5
1971	21.1		66.2	13.6	18.8	119.7
1972	20.5		66.3	12.4	18.6	117.8
1973[d]	19.3	3.6	59.3	13.2	17.7	112.1
1974	19.7	3.5	57.4	13.2	16.1	109.8
1975	20.4	3.6	62.1	13.8	20.4	120.1

Source: Department of Employment Gazette (various issues)
[a] Data refer to the United Kingdom for May of each year, and exclude marine engineering.
[b] Since 1973 unskilled production workers (mainly labourers) have been included in the otherwise semiskilled category 'other production workers'. For the purpose of this analysis these employees have been reallocated to 'other occupations' for 1964 and 1965.
[c] It is not possible to disaggregate 'other occupations' from 'other production occupations' for 1964 and 1965.
[d] In 1973 the occupational classification was revised to be compatible with the List of Key Occupations for Statistical Purposes with the effect that a new category – 'foremen and supervisors' – was introduced (see Department of Employment Gazette, September 1972, 799).

group higher in 1975 than in 1964 (unlike that of the other two main groups skilled and semi-skilled and unskilled occupations, which both fill) but the difference between the minimum and maximum level of employment and between 1964 and 1975 employment was greater than for any other occupational group and for total industry employment (table 5.2). Employment in skilled occupations (craftsmen, foremen and supervisors) was subject to much less variation, in aggregate, than was managerial and ATC employment. The figures for the latter are affected by steady growth throughout the period, and followed very closely the pattern for total employment in the industry. By contrast employment in semi-skilled and unskilled occupations showed least variation between 1964 and 1975 levels of employment, which hid considerable variations in the level of employment within the period (table 5.2).

Similar results have been obtained by Nissim (1984b) in a study of cyclical variations in employment in the British mechanical engineering industry in the period 1963–78, which is also based on data drawn from the Department of Employment's survey of occupations. Nissim identifies a differential cyclical response in the employment of different types of labour. In particular skilled craftsmen and non-manual workers fluctuate less than do those in the semi-skilled and unskilled categories. Following the work of Becker (1962), Oi (1962) and Rosen (1968) on the fixity and heterogeneity of labour, Nissim (1984b) attributes this differential to the existence of significant hiring, firing and training costs which effectively fix the stocks of some groups of workers in the short term. He does not, however, extend his discussion to examine the determinants of changes in aggregate which are reflected in this differential response.

These changes in aggregate occupational structure suggest two major consequences for labour demand in the shipbuilding industry. The first is the progressive growth in the relative and absolute importance of managerial and ATC employment – the substitution of non-production for production occupations – which is a general feature of the development of all manufacturing industries (Crum and Gudgin, 1977; Gershuny, 1978). The second is the tendency for employment in semi-skilled and unskilled occupations to decline more rapidly in both relative and absolute terms, and fluctuate more noticeably than employment in skilled occupations. The first of these two points reinforces the shift of employment into non-production occupations, as the skill requirement of the industry changes with changing process and product technologies. The second point reflects the tendency for certain occupational groups to face an inherently more unstable pattern of labour demand. This in turn reflects the combined influence of two factors. The short-run fixity of employment in skilled and nonmanual occupations is due

Table 5.2 Employment variation in the British shipbuilding industry, 1964–1975

Occupation	Minimum employment	Maximum employment	1964 employment	1975 employment	Maximum as per cent of minimum employment	1975 as per cent of 1964 employment
		(as per cent of total employment)				
Managerial and ATC	14.9	17.9	15.2	16.9	128.1	122.2
Skilled[a]	54.8	56.3	55.4	54.6	112.3	108.2
Semiskilled and unskilled[b]	26.3	29.6	29.4	28.4	119.8	105.9
Total	–	–	100.0	100.0	112.3	109.5

Source: Table 5.1
Notes: [a] Foremen, supervisors and craftsmen
[b] Other production and other occupations

to the existence of hiring, firing and training costs which lead firms to treat such employees as a quasi-fixed factor of production (Nissim, 1984b). Second there is the specific relationship between task and function, and hence occupational employment, and the shipbuilding production cycle. Apart from the managerial and ATC occupational group, in which a high maximum/minimum ratio reflects a fairly consistent growth in employment, it can be concluded from the evidence presented in this section that employment stability in the shipbuilding industry appears to increase with the skill level of the occupational group. The following section will examine in greater detail the pattern of employment stability for a more detailed disaggregation of occupations, and will relate this more closely to the nature and pattern of technical change in the industry.

OCCUPATIONAL ANALYSIS OF EMPLOYMENT STABILITY

The conclusion of the previous section, that employment instability increased as the skill level of shipbuilding employment fell, is confirmed from an analysis of employment change between 1966–72 (table 5.3). This indicates that employment in semi-skilled occupations fluctuated more noticeably than did employment in skilled crafts and declined by almost 9 per cent over the period, compared to a slight increase in the level of skilled employment. As a result, semi-skilled occupations accounted for 10.5 per cent of total industry employment in 1972 compared to 11.3 per cent in 1966.

Table 5.3 Employment variation in the shipbuilding industry, 1966–1972

	Maximum as per cent of minimum	1972 as a per cent of 1966
Skilled occupations	103.7	100.5
Semiskilled occupations	109.9	91.6
Total shipbuilding industry	104.5	99.0

Source: Department of Employment

However, more detailed examination indicates that there are important variations in employment stability within the skilled occupation group. Of particular interest, especially in view of the influence of the shipbuilding production cycle and product mix on the pattern of labour demand in the shipbuilding industry (Harrison, 1985), is the distinction between shipyard and outfitting trades. Employment in outfitting will tend to be less stable than employment in the shipyard trades for two reasons. First, a characteristic feature of the ship production cycle is that the labour demand curve for outfitting trades is much more steeply peaked than that for the shipyard trades. Employment in the latter rises rapidly on commencement of production and falls off progressively through the latter half of the construction period. This may extend over a 48 month period from initial design to post-delivery contractual obligations to the shipowners (Booz, Allen and Hamilton 1973). After launching, employment in the outfitting trades dominates, as these trades are put to work in strength to complete the ship. With low order books and low rates of capacity utilization, which have characterized the British shipbuilding industry in recent decades, the level and occupational structure of employment in the industry is consequently very responsive to changes in the level and composition of the industry's order book.

The second factor, changes in the product mix of the industry, reinforces the influence of the production cycle. Complete regularity and continuity of employment, both in total and at the level of individual occupational groups, is difficult to achieve in an industry with a large product produced in limited numbers, frequently to bespoke designs, and with an occupational labour input demand which varies over the production cycle. The changing pattern of demand for various types of ship itself necessitates changes in pattern of labour input. This is seen most clearly in the growing importance of oil tanker and bulk ore carrier construction and the growth in the average size of ships constructed to very simple designs (Todd, 1983; Harrison, 1983). The latter has greatly reduced the demand for outfitting trades compared to that generated by the construction of more specialized ship types such as passenger liners.

The influence of the production cycle suggests that the variability of employment in outfitting trades, as measured by the ratio between minimum and maximum employment levels, should be greater than for the shipyard trades. The influence of a changing product mix, towards the production of larger and simpler ships, should be reflected in a more rapid decline in outfitting employment over the period as a whole. It is clear [from table 5.4] that between 1963 and 1972 employment in shipyard trades was indeed less volatile than employment in outfitting trades (as indicated by the maximum/minimum ratio) and while employment in shipyard trades rose by about four per cent over the period employment in the outfitting trades fell by 15 per cent.

At a more disaggregated level still, there are considerable variations in employment stability among specific occupations (table 5.5). As expected,

Table 5.4 Employment variation in shipyard and outfitting trades in the United Kingdom, 1963–1972[a]

Occupation	Maximum as per cent of minimum employment	1972 as a per cent of 1963 employment
Shipyard	111.8	104.2
Outfitting	117.7	85.0
Total skilled	112.5	103.8
Total shipbuilding	112.2	102.4

Source: Department of Employment
[a] For a definition of the occupation composition of shipyard and outfitting trades see table 5.5

Table 5.5 Employment variation in the shipbuilding industry, by occupation, 1963–1972

Occupation	Maximum as per cent of minimum	1972 as per cent of 1963 employment
Shipyard trades		
Riveters	265.1	41.9
Holders on	268.4	37.3
Drillers	123.5	83.0
Ship and blacksmiths	126.3	73.2
Caulkers	126.4	119.2
Turners	126.5	94.6
Platers	129.7	120.9
Welders	127.1	126.5
Burners	154.3	154.3
Shipwrights	120.1	89.8
Riggers	138.9	92.6
Outfitting trades:		
Joiners	133.9	74.7
Electricians	128.0	81.8
Painters and decorators	143.4	75.4
Plumbers	136.8	104.9
Sheetmetal workers/coppersmiths	138.9	106.0
Miscellaneous:		
Mechanics and fitters	120.0	94.3
Other craftsmen	158.3	83.8

Source: Department of Employment

outfitting occupations, on the whole, experienced greater fluctuations in employment, as measured by the maximum/minimum ratio. However, their employment performance over the period is mixed, with all trades except plumbers and sheetmetal workers/coppersmiths experiencing a decline in employment of between 20 and 25 per cent. There are two possible reasons for this anomaly. The first is definitional in nature: these two occupations are only partly outfitting trades and may not therefore fully reflect the general trend in outfitting employment. The second reason is more substantial. As already suggested, there was a significant change in the nature of the output of the British shipbuilding industry over this period. There was an increase in ship size and a growing specialization of at least some important sections of the industry on the construction of tankers and bulk carriers in which pipework installation constitutes an important part of total outfitting requirements.

Within the shipyard trades themselves, which as a group are the most stable in employment terms, variations in employment performance on either measure are much more pronounced. Rather than examine in detail changes in employment in every shipyard trade, however, it is more illuminating to follow through the occupational consequences of the major process innovation in the post-1945 British shipbuilding industry – the introduction of the welded hull to replace riveting.

The introduction of welding significantly affected the division and occupational structure of shipbuilding in quantitative terms. Reflecting the earlier argument that technical change is a continuous transformation process rather than a discrete event, the displacement of riveting by welding was a gradual process. In the early 1930s, welders represented only 0.3 per cent of the total shipbuilding industry workforce in Britain. By 1945 this proportion had risen sharply to 3.4 per cent of the total workforce, or 5.8 per cent of the skilled workforce of the industry. In the same year riveters accounted for 11 per cent of the skilled workforce. However, by 1950 welding employment exceeded riveting employment for the first time. As table 5.6 indicates, welders grew steadily in numbers after 1950, and by 1960 they accounted for around 13 per cent of the skilled workforce. By contrast, between 1950 and 1965 riveters fell from 7.5 per cent to less than 3 per cent of the skilled workforce in shipbuilding.

These changes were not reflected more widely in other major steel trades. In part this reflects an inherent flexibility, interchangeability and adaptability in some trades which facilitated a positive response to changes in task and function following a change in the techniques used in the industry. Caulkers, for example, were able to move fairly easily into welding-related operations, such as the preparation of butt welds. Similarily, there was a marginal

Table 5.6 Distribution of selected steel working trades as a percentage of the skilled workforce, 1950–1965

Trade	1950	1955	1960	1965
Welders	8.7	10.7	13.1	13.0
Riveters	7.5	6.3	4.0	2.9
Platers	9.1	9.9	9.8	9.0
Caulkers	3.5	3.6	3.8	3.2
Burners	2.1	2.3	2.6	2.8
Shipwrights	11.5	12.1	11.7	13.1

Source: Shipbuilders Employers Federation Statistics quoted in McGoldrick (1984, 210)

improvement in the position of burners as a result of the widespread use of oxy-gas cutting and planing machine processes, which was later to become a contentious issue between burners and platers (McGoldrick, 1984). Within the industry it had initially been felt that the introduction of welding would lead to a significant reduction in the demand for platers. That this did not, in fact, occur was due to two factors: welding, like pneumatic and hydraulic riveting before it, was eventually introduced as a negotiated rather than an imposed technical change; and platers found a new role in undertaking preparatory work for welders.

The implication of this brief case history is that the introduction of new processes, materials and machinery, which define tasks and functions and the manner in which they are to be carried out, necessitates a new division of labour. The need for traditional craft skills may decline, as experienced by riveters, or the content of these skills may change (e.g. caulking, plating), while new skills may arise (welding). One major consequence of this, which was noted in the Geddes Report (Shipbuilding Inquiry Committee 1967) and has been explored in detail elsewhere (McGoldrick, 1984), was the stimulation and intensification of demarcation disputes and an increased resistance to changes in traditional working practices.

The evidence from the Department of Employment occupational surveys shows that many of these trends continued into the 1970s. Between 1963 and 1972 employment in riveting and related trades (holders on and drillers in the main, but also in part shipsmiths and blacksmiths) fell dramatically – by around 60 per cent in the case of riveters and holders on. In addition, employment variation in these two trades during the period was almost twice as great as for any other single occupation. By contrast, employment in welding and related trades (such as burners) continued to increase sharply and showed much less variation than that in riveting-related trades (the exception is for burners, the high employment variability of which reflects continued growth, particularly after 1969, rather than fluctuations as such).

One further consequence of the adoption of welding in place of riveting during this period is seen in the relative expansion of employment in plating, which grew by almost 21 per cent. This reflects the growing need to prepare steel plates to a higher degree of accuracy when being abutted for welding rather than overlapped for riveting. It suggests that compared to the earlier period the trade-off between plating work for welding and that done for riveting was more straightforward.

It is clear from this analysis that there have been considerable variations in both the level and variability of employment in occupations in the ship-building industry which can be related back to changes in the product and process technology of the industry. These variations have been identified on the basis of a year to year analysis of the occupational structure of the British shipbuilding industry as a whole. They confirm the findings of previous studies of labour turnover in the shipbuilding industry which have relied on survey evidence from one (Robertson, 1954) or a small number (Hunter, 1967) of yards to identify week to week variations in the demand for labour.

In both earlier studies, it emerged that employment variations were greatest among the outfitting trades, whether measured by the range between maximum and minimum employment or labour turnover rates (Robertson, 1954) or by a stability index measuring the attachment of labour to the firm (Hunter, 1967). Aside from the influence of the production cycle and, to a lesser extent, product mix (which are particularly important influences on week to week variations in employment) this fundamental difference in stability and turnover rates may reflect two further factors. First, the finishing trades – plumbers, painters, joiners, electricians, and some sheetmetal workers – have considerable scope for employment in other industries, particularly but not exclusively the construction industry. In other words, their skills are not industry specific. Second, shipyard employers may have been more prepared to make the finishing trades redundant than the shipyard (or 'black') trades. This may be a tacit admission by the employer that these trades may be able to get alternative employment more easily, or it may be an acknowledgement that such labour can be attracted back to the shipyard when needed. Certainly from the evidence presented by Hunter (1967) for the Upper Reaches of the Clyde, which is confirmed by studies of shipbuilding redundancies in other areas (Sams and Simpson, 1968; Mackay et. al., 1980), redundancy among the finishing trades does seem to have been higher than in the shipyard trades.

REGIONAL VARIATIONS IN OCCUPATIONAL STRUCTURE

These temporal variations in the occupational structure of shipbuilding employment are accompanied by regional variations due to the different combined influence of the various factors, such as product mix and the production cycle, which influence the industry's demand for labour (Harrison, 1985). This section examines the evidence for these regional variations, using economic activity data from the 1971 Census of Population (OPCS, 1975; Northern Ireland General Register Office, 1977). The results of the 1981 Census of Population are not used, because the economic activity data have been prepared on the basis of the 1980 Standard Industrial Classification (SIC), rather than the 1968 SIC used here, and do not give data for an equivalent industrial group.

At the outset it can be noted that variations in employment status in 1971 were considerable, reflecting interregional variations in the structure of the shipbuilding and marine engineering industry (table 5.7). Two groups of regions can be identified. On the one hand are the traditional shipbuilding regions (such as Scotland, Northern Ireland, the Northern region, and to a lesser extent, the North West) which are characterized by low levels of self employment (less than half the national average) and managerial employment and a higher than average representation of other employees. These reflect the domination of the industry here by large shipbuilding and marine engineering works. By contrast, in the second group of regions, such as the South East, which are characterized by a more diversified industry structure, self employment and managerial employment are more important than the national average.

Of greater interest, however, are regional variations in the structure of occupations within the shipbuilding and marine engineering industry. The 10 per cent sample economic activity tables from the 1971 Census of Population (OPCS 1975, table 21) identify nineteen occupation unit groups which each contain 250 or more persons. Comparable employment data for Northern Ireland have been extracted from the Northern Ireland Economic Activity tables (Northern Ireland General Register Office, 1977, table 11). As expected, there are considerable interregional differences in the relative importance of these occupational groups. The nineteen occupations account for between 77 per cent and 85 per cent of total male employment in the industry in these regions, and this proportion tends to be highest in the traditional shipbuilding regions. Within each region the importance of individual occupations varies considerably – only one of the nineteen groups (warehousemen, storekeepers and assistants) does not appear among the ten most important occupations in any region.

Table 5.7 Shipbuilding and marine engineering: persons in employment by status, 1971

Region	Self-employed	Managers	Foremen and supervisors	Other	Total Employment
		(as a percentage of total employment)			
Northern Ireland	0.31	1.43	5.01	93.25	9654
Scotland	0.44	2.22	3.72	93.63	40630
Northern	0.21	1.81	4.23	93.75	37610
North West	0.29	2.50	4.99	92.23	27660
Yorkshire and Humberside	1.01	4.46	5.61	88.92	6950
South East	1.48	4.99	4.58	88.87	41310
South West	0.98	2.62	3.05	93.35	16400
United Kingdom	0.79	3.05	4.34	91.81	190074

Source: OPCS (1975, table 21); Northern Ireland General Register Office (1977, table 12)

Rather than look in detail at interregional variations in the occupational structure of the shipbuilding and marine engineering industry, it is useful to consider the impact of technical change on regional occupational structures. Reference has already been made to the progressive replacement of riveting by welding since 1945, and the effect this has had on national employment trends in these and related occupations. Differences in product type and production technology between regions will be expected to issue in similar differences in the relative importance of occupation unit groups among regions. From an examination of employment in metal-working occupations (table 5.8) it is clear that while there is a marked distinction between Northern Ireland, Scotland and the Northern region and the other regions (with the exception of Yorkshire and Humberside) in the overall importance of the metal-working trades, there are also significant interregional differences in occupational balance within the metal-working trades.[1] This can be demonstrated most clearly by reference to Northern Ireland, where the industry is dominated by a single large company (Steed, 1968; Harrison, 1985). Accordingly, the attribution of observed differences in occupational composition to structural influences, particularly the underlying technology of production reflected in the tasks and functions performed in the work-place, is greatly facilitated. Table 5.8 makes it clear that welders are of approximately equal importance in Northern Ireland, Scotland and the Northern region, where they account for a higher proportion of total industry employment than elsewhere. However, although figures for the employment of riveters are not separately available, table 5.8 also shows that in Northern Ireland metal plate workers and riveters as a group are much less important compared to the other two major shipbuilding regions. Other metal-working occupations (notably steel erectors, riggers, sheet metal workers, and other metal making) are correspondingly more important. This reflects the nature of the shipbuilding industry in Northern Ireland, which by 1971 was specializing to a greater extent than other British ship-yards in the construction of very large and ultra large crude oil tankers and bulk carriers in which the welding of steel plates constituted a more significant element in total shipyard activity than in the more diversified industries of the other regions (Harrison, 1985). The different combination of tasks and functions which characterize the technology of the Northern Ireland shipbuilding industry, when compared with the less specialized nature of the industry in other regions, is reflected in the fact that in 1971 the ratio of metal plate workers and riveters to welders was lower in Northern Ireland than in any other region (table 5.9). In Northern Ireland there were 60 welders to every 100 metal plate workers and riveters. This is 50 per cent higher than the national industry average and between 6 and

Table 5.8 Employment in metal-working occupations in shipbuilding and marine engineering (males), 1971

Occupation	% total male workforce							
	NI	SC	N	NW	YH	DR	DE	UK
033 Sheet metal workers	3.3	2.2	0.9	2.9	1.8	1.6	2.1	1.9
034 Steel erectors; riggers	5.3	1.6	1.7	2.8	0.9	2.3	2.6	2.1
035 Metal plate workers; riveters	13.3	18.4	19.4	13.1	13.9	13.3	16.5	15.7
036 Gas, electric welders, cutters; braziers	8.0	8.0	10.4	6.8	4.1	2.6	7.1	6.5
054 Other metal making, working	7.8	4.9	5.2	3.8	2.4	3.2	9.9	4.5
Total	37.6	35.2	37.5	29.4	20.0	22.8	38.2	30.6

Source: OPCS (1975, table 21); Northern Ireland General Register Office (1977, table 11)

Table 5.9 Regional variations in the employment of welders and metal plate workers and riveters, 1971

Region	No of welders per 100 metal plate workers and riveters
Northern Ireland	60
Scotland	44
Northern	54
North West	52
Yorkshire and Humberside	43
South East	19
South West	29
United Kingdom	41

Source: As for table 5.8

16 percentage points higher than in the other three major shipbuilding regions in Great Britain.

CONCLUSION

If, as was argued at the start of this chapter, the process of production can be understood functionally to comprise the performance of a series of inter-related tasks, and a technique is defined as a combination of tasks to be performed to achieve a particular result, technical change can be thought of as a change in the range of functions to be performed and the methods used to perform them. Technical change is a continuing transformation process in an economic system (which may be a sector, enterprise or establishment) rather than a discrete event or jump between two states. Accordingly, the transition path from the old to the new technology is characterized by the creation, alteration and obsolescence of tasks, functions and their dependent occupations.

Some aspects of the transition path have been explored in the context of occupational changes in the UK shipbuilding industry in the 1960s and 1970s. Although the data are less than comprehensive, this chapter has established that there have been important temporal and spatial variations in the industry's occupational structure over the study period which can be understood in terms of the ongoing process of technical change in and affecting the industry. Some of these changes, such as the general drift towards a higher proportion of the workforce in managerial and administrative, technical and clerical occupations, reflect widespread changes in the occupational structure of manufacturing as a whole. Other changes, notably reflected in differences in the variability of employment in different occupational groups, reflect the combined effects of changing product and process technologies in the industry.

Employment in the shipyard trades, for example, has been shown to be less variable than employment in outfitting trades, confirming at the macro level the results of earlier microlevel case study work based on a week to week analysis of labour turnover in specific shipyards. This has been traced back to the influence of the shipbuilding production cycle on the pattern of labour input demand for these trade groups and to the impact of changes in the product mix of major sections of the British merchant shipbuilding industry. Within the shipyard trades themselves different occupations have experienced different rates of employment change. In particular, the introduction of the welded hull, the major process innovation in the post-1945 shipbuilding industry, has had a dramatic impact on the occupational structure of and division of labour in the industry. Over a forty year period this particular process innovation has redefined the tasks and functions which constitute shipbuilding production in Britain, with the result that some traditional craft skills have become increasingly obsolete: the nature, content and context of other occupational skills has changed and new skills have arisen.

The specific relationship between technology and changes in occupational structure, and its implication for the national and regional development of the shipbuilding industry, constitutes one important part of the assessment of the overall labour market implications of major industrial change and restructuring in the context of the UK shipbuilding industry (Harrison, 1985). More generally, changes in occupational structure are an important aspect of the wide structural reorganisation of production in the UK (Massey, 1984). Accordingly, further studies of the specific technology/occupation-change relationship in specific sectors can usefully be integrated with the wider debate which emphasizes the role of job change, the redivision of labour and restructuring of skills on the process of class restructuring and the restructuring of the industrial space economy in advanced capitalist societies (Bradbury, 1985).

NOTES

1 In this discussion and the remainder of this section Wales, East Anglia, West Midlands and East Midlands have been omitted from the analysis due to the small relative and absolute size of the marine industry in these regions.

REFERENCES

ACARD (1978) *Industrial Innovation*, HMSO, London.

Albu, A. (1976) Causes of the decline in British merchant ship-building and marine engineering, *Omega* 4, 513–25.

Albu, A. (1980) Merchant shipbuilding and marine engineering. In Pavitt K. (Ed.) *Technical Innovation and British Economic Performance* Macmillan, London, 168–83.

Al-Timimi, W. (1976) Innovation led expansion: The shipbuilding case, *Research Policy* 4, 160–71.

Becker, G. S. (1962) Investment in human capital: a theoretical analysis, *Journal of Political Economy* 70 (Supplement) 9–49.

Booz, Allen and Hamilton (1973) *British Shipbuilding 1972: A report to the Department of Trade and Industry*, HMSO, London.

Bradbury, J. H. (1985) Regional and industrial restructuring processes in the new international division of labour, *Progress in Human Geography* 9, 38–63.

Brown, R. and Brennan, P. (1970a) Social relations and social perspectives amongst shipbuilding workers – a preliminary statement. Part I, *Sociology* 4, 71–84.

Brown, R. and Brennan, P. (1970b) Social relations and social perspectives amongst shipbuilding workers – a preliminary statement. Part II *Sociology* 4, 197–211.

Crum, R. E. and Gudgin, G. (1977) *Non-production Activities in UK Manufacturing Industry*. Regional Policy series 3, Collection Studies, Commission of the European Communities, Brussels.

Gershuny, J. I. (1978) *After Industrial Society?* Macmillan, London.

Henwood, F. (1984) *Science, Technology and Innovation: a Research Bibliography*, Wheatsheaf Books, Brighton.

Harrison, R. T. (1983) Consequences of technological change: the case of the shipbuilding industry. In Bosworth, D. (Ed.) *The employment Consequences of Technological Change*, Macmillan, London, 157–73.

Harrison, R. T. (1985) The labour market impact of industrial decline and restructuring: the example of the Northern Ireland shipbuilding industry, *Tijdschrift voor Economische en Sociale Geografie* 76, 331–44.

Henwood, F. (1984) *Science, Technology and Innovation: a Research Bibliography*, Wheatsheaf Books, Brighton.

Hogwood, B. (1979) *Government and Shipbuilding. The Politics of Industrial Change*, Saxon House, Farnborough.

Hussain, A. (1983) Theoretical approaches to the effects of technical change on unemployment. In Bosworth, D. (Ed.) *The Employment Consequences of Technological Change*, Macmillan, London, 13–24.

Hunter, L. C. (1967) *Problems of labour supply in shipbuilding*. Second Marlow (Scotland) Lecture. The Institution of Engineers and Shipbuilders in Scotland, Glasgow.

McGoldrick, J. (1984) Industrial relations and the division of labour in the shipbuilding industry since the war *British Journal of Industrial Relations* 21, 197–220.

98 *Harrison*

Mackay D., Mackay R., McVean R. and Edwards R. (1980) *Redundancy and Displacement*. Research Paper 16, Department of Employment, London.

Massey, D. B. (1984) *Spatial Divisions of Labour: Social Structures and the Geography of Production*, Macmillan, London.

Nissim, J. (1984a) The price responsiveness of the demand for labour by skill: British mechanical engineering: 1963–1978, *Economic Journal* 94, 812–825.

Nissim, J. (1984b) An examination of the differential patterns in the cyclical behaviour of the employment, hours and wages of labour of different skills: British mechanical engineering, 1963–1978 *Economica* 51, 423–36.

Northern Ireland General Register Office (1977) *Census of Population 1971: Economic Activity Tables Northern Ireland*, HMSO, Belfast.

Oi, W. (1962) Labour as a quasi-fixed factor of production *Journal of Political Economy* 70, 538–555.

OPCS (1975) *Census 1971, Great Britain. Economic Activity Part III (10% sample)*, HMSO, London.

Robertson, D. J. (1954) Labour turnover in the Clyde shipbuilding industry *Scottish Journal of Political Economy* 1, 9–32.

Rosen, S. (1968) Short-run employment variations on class I railroads in the US: 1947–1963 *Econometrica* 36, 511–29.

Rosenberg, N. (1976) *Perspectives on Technology*, CUP, Cambridge.

Sams, K. I. and Simpson, J. V. (1968) A case study of a shipbuilding redundancy in Northern Ireland, *Scottish Journal of Political Economy* 15, 267–82.

Shipbuilding Inquiry Committee (1967) *Shipbuilding Inquiry Committee 1965–1966 Report*. Cmnd 2937, HMSO, London.

Steed, G. P. F. (1968) The changing milieu of a firm: a case study of a shipbuilding concern, *Annals, Association of American Geographers* 58, 506–25.

Todd, D. (1983) Technological change, industrial evolution and regional repercussions: the case of British shipbuilding, *Canadian Geographer* 27, 345–60.

Venus, J. (1972) The Economics of Shipbuilding, Sixth Blackadder Lecture, University of Newcastle-upon-Tyne.

PART II

Technical Change and Industrial Restructuring

6

Innovation, adaptation and survival in the West European oil refining industry

D. A. Pinder and M. S. Husain

New high-technology industries have recently attracted much publicity, but technological change is also central to the development of mature industries in adjusting to their changing economic environments. Emerging disadvantages may be overcome and new opportunities may be seized. But many investigations have demonstrated that by no means all firms in a mature industry take the same stance with respect to technological change. Some adopt offensive strategies (Malecki, 1980, p. 222). This group naturally includes leaders who invest, often heavily, in research and development (R&D) for new technologies. Others lead in a different sense: by identifying and assimilating advantageous technologies from other industries. Computers and computer-controlled production methods are good examples (Gibbs and Edwards, 1985; Rees, Briggs and Hicks, 1985). A third group opts for, or perhaps through lack of resources is forced into, imitative strategies: the adoption of technological developments achieved by other firms in the industry. Lastly, of course, there may well be a fourth group, firms which do not adopt particular innovations, either through choice or through lack of awareness. This fourfold division provides an approach to the problem of clarifying the complex currents of technological change in mature industries, but it is also helpful to view these currents from other perspectives.

First, irrespective of whether a strategy is offensive or imitative, it is necessary to reiterate the well-organized distinction between the two main goals of technological change: product innovation and process innovation. There is a fundamental difference between those firms and industries needing to move on to new products (Le Heron, 1980; Thomas, 1985, p. 26–7) and those requiring new processes to enhance the comparative advantage of existing product lines (Gibbs and Edwards, 1985).

Second, in the case of multiplant firms it is necessary to distinguish between strategies for technological change at the level of the firm and those at the level of the plant. A firm developing or adopting a technology may not choose to diffuse it throughout the corporation, even if all plants are producing essentially the same product(s). Such intracorporate choices may be logical and may in reality pose no threat to the 'neglected' locations. Yet these decisions prompt questions concerning the future of plants not selected for investment, especially if they are taken in an era of economic difficulty.

Third, insights can be gained by taking a theoretical perspective on long-term, intracorporate technological planning. Particularly relevant is Etzioni's (1973) 'mixed-scanning' theory of planning. This postulates an alternation between lengthy periods in which incremental planning is dominant, and shorter episodes of far-reaching strategy reappraisal stimulated by crises. Applied to the technological evolution of many mature industries, this theory would interpret the thirty years up to the mid-1970s as an incremental planning era. During this period corporate technological strategies evolved steadily as new technologies opened up new product or process opportunities, or as shifts in the economic environment induced planners to select a technological response. Subsequently, economic crisis has thrust on to many firms the need to review much more radically their medium- and long-term plans for product and process development. For firms able to achieve successful strategy reorientation, it is likely that incremental planning will once more reassert itself and will be sufficient to maintain firms on satisfactory development trajectories for an unspecified period.

Fourth, this theoretical standpoint should be related to the timescales needed to pursue offensive and imitative strategies, because these timescales are often markedly different. R&D fundamental to an offensive strategy may take a decade or more to complete, whereas the adoption of a known technology rarely spans more than four or five years, even when extensive re-equipment or construction are required. When industries face sudden major crises, therefore, imitative strategies may well be the most appropriate short-term response since they offer tried and tested solutions within predictable time and cost limits. For firms with sufficient resources, further improvement through R&D may then become a long-term goal.

OIL REFINING AND TECHNOLOGICAL CHANGE

Oil refining was one of the most impressive industries to reach maturity in

Europe in the post-1945 era. Total refining capacity rose from a mere 41 m tonnes in 1950 to 1034 m tonnes in 1976.[1] This reflected unprecedented demand growth, but successful expansion was also founded on technological advances. In the case of process technologies, R&D allowed more precise refining of fuels and the opportunity to benefit from major economies of scale (Molle and Wever, 1984a, p. 52–5). Large-scale refining was also made feasible by the development of automated control processes, essential in the safety field. Product innovations, meanwhile, were not particularly evident in basic fuels, but were impressive in the downstream petrochemicals industry.

As the refining industry expanded, company planning showed strong incremental tendencies. Larger markets meant that corporate stategies were adjusted to build new, more advanced, refineries and expand existing ones. Similarly, periodic maintenance programmes provided the opportunity for improved technologies to be added to refineries, if they were perceived to be commercially justified. But this incremental process was severely checked by the first oil price crisis of 1973–4 and was clearly ended by the second crisis of 1979–80. As mixed-scanning theory predicts, these crises forced the industry into unprecedented financial losses necessitating urgent revision of long-term company planning strategies. Two problems – overcapacity and production inflexibility – were responsible for the industry's difficulties. Overcapacity was tackled by abandoning expansion schemes and substituting far-reaching refinery closure and contraction programmes. Briefly, crude oil refining capacity in Western Europe was cut by 2 per cent between 1976 and 1979, and by 1985 was reduced by a quarter (Pinder, 1986). Inflexibility, in contrast, generated a totally different response: heavy investment in technologies enabling the industry to adjust its product profile much more closely to the changing demand profile of the European market.

Inflexibility, i.e. the restricted ability to maintain output of high-demand products whilst sharply curtailing the output of products for which demand is weak, developed by default during the post-1945 expansion phase.[2] During this period R&D made available and improved the so-called conversion technologies. These offered flexibility because of their potential for transforming heavy refinery products into lighter fractions. But the adoption of these technologies was far from complete, even in the refining systems of major companies that were heavily involved in R&D. While the market remained buoyant, the need for conversion facilities often did not seem central; what could be produced could be sold. But after the first oil price crisis new demand trends became established: in particular, while gasoline and middle-distillate consumption held up relatively well, sales of residual fuel oil rapidly declined (table 6.1). Fuel-oil surpluses developed, and the scramble for conversion technologies was triggered.

Table 6.1 Changes in demand structure, 1973–1984

	Percentage change		
	1973	*1984*	*Demand change (%)*
Gasolines	17.7	23.2	+3.7
Middle distillates	33.2	37.1	−11.9
Fuel oil	36.0	24.0	−47.4
Other products	13.1	15.7	−5.2
	100.0	100.0	−21.1

	Absolute change, 1979–84 (m tonnes)						
	1973	*1979*	*1980*	*1981*	*1982*	*1983*	*1984*
Gasolines	132.6	148.7	142.4	131.0	134.2	137.0	137.5
Middle distillates	248.6	254.6	234.0	223.7	215.6	218.0	219.0
Fuel oil	269.9	231.8	211.7	187.0	165.4	142.6	141.9
Other products	97.7	97.3	92.0	90.0	89.1	88.1	92.6
	748.9	732.4	680.1	632.6	604.3	585.7	591.0

Sources: British Petroleum (1984), 10 and (1985), 10

In some cases, notably involving major companies, offensive strategies were adopted. Several large-scale R&D programmes are now in progress to devise new processes able to deal with intractable residues left at the end of existing refining processes. In their most extreme form these programmes are aiming for total conversion of fuel-oil residues. Progress is being made (Bergren, 1984), yet the R&D timescale is such that the impact of new process technologies was negligible by 1985. Meanwhile, however, imitative strategies have also been pursued by many large and small refiners who have reasoned that existing operations involving known conversion technologies are well placed to adjust to the immediate crisis. The use of known technologies has expanded impressively, especially since 1979, and there is every sign that they will dominate the process of technological change to the end of the 1980s. In this investigation, therefore, the focus is on these process-oriented imitative strategies, looking particularly at projects completed since 1979. Before the results are examined, however, the technologies themselves must be considered.

CONVERSION TECHNOLOGIES

Basic refining employs atmospheric distillation processes. Simplifying greatly, crude oil is heated at atmospheric pressure – a process requiring relatively simple equipment – so that the resulting distillates may be collected in a fractionating tower. For a typical crude oil this process may well result in 45–50 per cent of output being residual fuel oil, a high proportion which becomes the target of conversion processes.[3] Extensive accounts of these processes have been provided by Plummer (1973), Haines (1978) and Royal Dutch Shell (1983). Here we focus on central features of the available technologies, on their relationships and on their relative effectiveness. This complements other summaries provided by Molle and Wever (1984a, 1984b).

Thermal operations

One relatively easy route to a significant improvement in a refinery's product profile is investment in downstream thermal operations (figure 6.1). In this case the dominant technologies are thermal cracking and visbreaking. With thermal cracking, residual fuel oil from the basic atmospheric distillation process is subjected to high temperatures (c. 530°C) and pressures (up to 13 atmospheres). Under these conditions many of the lengthy, heavy hydrocarbon molecules that comprise fuel oil 'crack' to release lighter molecules and, therefore, products. Although conversion is never complete, this process is capable of reducing the fuel-oil yield of a typical crude oil from almost 50 per cent to little more than 30 per cent (Molle and Wever, 1984a, p. 13–14). The effectiveness of this process should not be overemphasized, however, since the improvement is split between middle distillates and gasolines, rather than being concentrated on gasolines alone (figure 6.1). On this criterion, thermal cracking is generally considered to be only 65 per cent as effective as the more advanced techniques (Commission of the European Communities, 1985, p. 25).

Visbreaking (literally viscosity breaking) differs from thermal cracking in that the fuel-oil feedstock is subject to high temperatures and pressures for a shorter time. This lowers conversion levels: on average, visbreakers are only one third as effective as catalytic crackers, but this disadvantage must be weighed against their attractions. Visbreakers are easily and cheaply added to an existing refinery. It is not unknown, for example, for them to be manufactured from redundant atmospheric distillation units. Also, because process times are shorter, costs are lower; and the yield of unwanted products, particularly furnace coke, is generally significantly lower than in full thermal-cracking units.

Vacuum distillation

In this process residual fuel oil from atmospheric distillation units is heated, vaporized and fed to a vacuum column, together with superheated steam

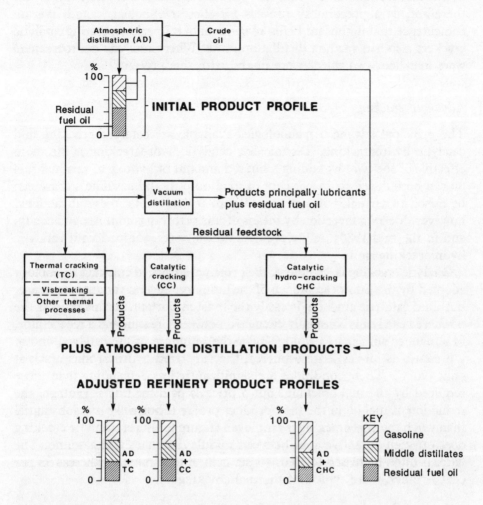

Figure 6.1 Conversion technologies and the adjustment of refinery output

to intensify the vacuum. The aim is to separate distillates which at atmospheric pressure would require temperatures leading to undesirable disintegration of the hydrocarbon molecules. Largely because vacuum distillation aims to prevent unwanted hydrocarbon division, it is not normally considered in the context of fuel-oil conversion. For two reasons, however, the technique has been included in this investigation. First, vacuum distillation does assist the improvement of the product profile, particularly by the production of distillates suitable for lubricant blending. Second, fuel oil treated by this process is more suitable as a feedstock for catalytic crackers and catalytic hydrocrackers than is raw fuel oil. Vacuum distillation is often employed, therefore, as a preparatory process for these technologies, and it is no coincidence that in the late 1970s 80 per cent of refineries operating catalytic crackers also had vacuum distillation units. Where catalytic hydrocrackers were installed, two thirds were linked with this process.

Advanced cracking

The principal advanced technologies available are catalytic cracking and catalytic hydrocracking. Technically, catalytic hydrocracking is the more effective of the two: by adding a limited amount of hydrogen, residual fuel oil can be reduced to 25 per cent of total output, and gasoline's share can be raised to virtually 50 per cent (figure 6.1). Despite these attractions, however, 'deep' conversion by means of this pathway is complex and costly, and in the mid-1980s only eleven refineries were equipped with catalytic hydrocracking units.

Catalytic cracking is, therefore, the principal advanced cracking technology adopted by the European branch of the industry. Several variants are in use, but fluid catalytic cracking is easily the most important. In this process the powdered catalyst is constantly circulated between a reactor and a regenerator, in a manner similar to a fluid. Under the influence of the catalyst, and at a pressure as low as 2 atmospheres, the vaporized hydrocarbons crack at only 490°C.[4] These conditions are significantly less demanding than those required by thermal cracking, but a price is paid for this advantage: the reduction of fuel oil in the final product profile is normally less substantial than when thermal cracking is employed (figure 6.1). Yet catalytic cracking does offer an attractive split between middle distillates and gasoline. The latter is often increased to 38 or 40 per cent of total output, whereas 33 per cent is much more typical of thermal cracking.

ASPATIAL AND SPATIAL TRENDS, 1979–85

Data detailing the adoption of individual processes, and combinations of processes, clearly demonstrate that in the late 1970s the adoption of conversion technologies had made only limited progress in Western Europe (table 6.2). This was despite the industry's popular reputation for technological advance. It is true that, in terms of individual processes, 44 catalytic crackers were in operation, as were nine catalytic hydrocrackers. Also, with respect to process combinations, 44 refineries had two conversion technologies on stream, and in nine cases three technologies were in use (figure 6.2). Yet

Table 6.2 Change in the application of technologies, 1979–1985

Operational 1979	Change caused by by closures	Tech- nologies surviving	Change caused by investment	Operational 1985
Refineries with:				
no downstream conver- sion technology	45	− 10	35	− 17	18
vacuum distillation	90	− 14	76	+ 6	82
thermal operations	29	− 2	27	+ 32	59
catalytic cracking	44	− 2	42	+ 13	55
catalytic hydro-cracking	9	− 1	8	+ 3	11

Source: SUORD

these aspects of the industry's technological profile must be set against its overall scale: out of 156 refineries operating in 1979, only 34 per cent were equipped with advanced cracking facilities or were operating two or more conversion technologies. Conversely 38 per cent operated only one conversion process in 1979, while a further 28 per cent had no conversion capacity and therefore relied on atmospheric distillation to determine the product split. Moreover, in the 60 refineries that had one process on stream, the dominant technology (found in 43 cases) was vacuum distillation. As the earlier technical discussion indicated, in itself vacuum distillation has very limited conversion power.

Subsequent developments brought a radical change in this bias towards low-conversion refineries (table 6.2). As is now well known, one factor contributing to the improved balance between simple and more advanced refineries has been the closure movement (Commission of the European Communities, 1985; Molle and Wever, 1984a, 1984b; Pinder, 1986) Thirty one refineries ceased production during the period. Although this led to the loss of a limited number of advanced conversion units, closures led primarily to the loss of vacuum distillation facilities and refineries with no conversion capacity.

Simply by adopting discriminatory disinvestment strategies, therefore, refiners brought about a significant adjustment in the overall technological profile. But this process should not be allowed to obscure the reshaping achieved by investment. One major aspect of this was to add a downstream process to refineries which previously relied solely on atmospheric distillation. This upgrading did more than the closures to reduce the number of very simple refineries from 45 to 18 (table 6.2). Simultaneously, however, many refineries which previously operated at least one conversion technology were also upgraded. One result of this was that refineries operating two

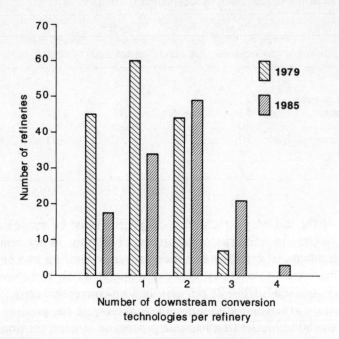

Figure 6.2   Conversion technology profiles for the industry, 1979 and 1985

conversion processes became the modal type (figure 6.2). This group accounted for 39 per cent of all refineries in 1985, the principal technological combinations being vacuum distillation coupled with either catalytic cracking (26 cases) or thermal operations (15 cases). In addition, a significant increase occurred in the number of refineries operating three or more conversion processes. Their proportional importance rose from 4.5 to 18 per cent, the dominant combination in this instance being vacuum distillation plus catalytic cracking and thermal operations (18 cases).

Taking a broad view therefore, a sharp distinction is evident between the industry's technological profiles at either end of this period, despite its relative brevity. Initially, two thirds of all refineries either had no conversion capacity or were limited to one process, typically vacuum distillation. By the end, nearly 60 per cent had two or more processes on stream. This shift was in part a reflection of the closure process, but this caused the loss of only 19 vacuum distillation, thermal or catalytic conversion units, compared with 54 such units commissioned during the period.

It has been shown elsewhere that the crisis in oil refining has been felt throughout the European system, and spatial aspects of the technological response must therefore be considered (Pinder, 1986). Has investment led to an essentially uniform technological adoption surface, or can significant variations be identified? This issue will be examined in three contrasting spatial contexts: the production systems of major firms, the location patterns of simple and complex refineries and, finally, the north-south balance of conversion capacity in Europe.

Oil majors: intracorporate investment strategies

Five major international companies, BP, Shell, Esso, Mobil and Texaco, operated 53 European refineries in 1979. This was a third of the total, and they accounted for 39 per cent of refining capacity. Two other majors, Caltex and Gulf, were also present in the late 1970s but have been excluded from the analysis because of their subsequent virtual withdrawal from European refining.

Partial withdrawal was, of course, an element in the restructuring strategies of the remaining majors. All of them closed refineries (table 6.3), the most significant point being that seven of the eleven abandoned installations had no conversion capacity more effective than vacuum distillation. Indeed, three had no conversion capacity at all. Closures primarily affected simple refineries, therefore, the result being a significant reduction in intracorporate technological contrasts.

Table 6.3 Process expansions and additions by major companies, 1979–1985

	Refineries operating		No. of process expansions in individual refineries	No. of process additions to individual refineries
	1979	1985		
Shell	16	15	—	3 (3)
BP	12	8	—	5 (3)
Esso	15	12	6 (6)	1 (1)
Mobil	7	5	4 (3)	1 (1)
Texaco	3	2	1 (1)	3 (2)
Total	53	42	11(10)	13(10)

In some instances individual refineries received more than one process expansion or addition. Bracketed figures indicate the number of refineries affected.
Source: SUORD

Beyond this, the process of eliminating the most rudimentary elements in company systems was sustained as the majors invested to upgrade surviving refineries. In terms of projects completed, upgrading i.e. the addition of processes to refineries, was slightly more important than the expansion of processes already on stream (table 6.3). Most significant of all, eight of the ten refineries affected by upgrading were originally equipped with no conversion facilities or, at best, with vacuum distillation. The outstanding example of this was the BP refinery at Rotterdam, one of the world's largest, which at the outset had no conversion capacity. It appears therefore that the general aim of corporate investment was not to transform existing advanced refineries into even more complex ones, but to restructure those with particularly narrow technological bases. This strategy led to a further reduction of technological contrasts between refineries. It also raised the issue of how simple refineries should be selected for upgrading or, at worst, closure. While each case was undoubtedly complex, what is evident is that size was a major criterion. The average crude capacity of simple refineries selected for closure by the majors was 5 million tonnes, but those selected for upgrading averaged 10.7 million tonnes.

Three further features of the majors' policies must be noted. First, although all companies employed the upgrading strategy, their reliance on it was not uniform. This was largely a reflection of earlier investment policies. The BP, Shell and Texaco systems all included a limited number of substantial simple refineries which came to require upgrading, whereas Esso and Mobil had previously distributed conversion technologies more widely. This enabled Esso and Mobil to extend their conversion capabilities by the expansion of processes already on stream (table 6.3).

Second, although intracorporate technological contrasts were muted by the closure or upgrading of simple refineries, the latter were not completely eliminated. Of the 18 refineries which survived without conversion facilities in 1985, three were owned by Shell, one by BP and one by Esso. The largest of these was BP's 5.2 million tonne Gothenberg refinery, and the remainder were substantially smaller. This retention of small, extremely simple refineries in contradiction of a well marked general trend clearly requires further investigation.

Third, because of their prominence and their financial resources, it might be assumed that the majors dominated the investment wave. The evidence does not support this. Although the majors increased their share of catalytic hydrocracking between 1979 and 1985, their shares of the much more popular catalytic cracking and thermal processes decreased markedly (table 6.4). The high level of technological adoption previously displayed by the majors was therefore emulated by many smaller companies. This reduction in technological contrasts between major refiners and the remainder of the industry implies the spatial spread of technological enhancement to localities and areas in which major companies were not strongly represented. This highlights the importance of further analyses at the international scale. Do

Table 6.4 Major company shares of European capacities, 1979–85

| | Company capacity (000 barrels per day) | | Percentage share | |
	1979	1985	1979	1985
Atmospheric distillation	7714	5836	38.5	32.5
Vacuum distillation	1012	1108	34.4	32.5
Thermal operations	392	527	55.0	32.4
Catalytic cracking	529	668	54.3	43.4
Catalytic hydro-cracking	43	98	33.3	50.3

Companies included are Shell, BP, Esso, Texaco and Mobil.
Source: SUORD

these confirm that the technological surface is now relatively uniform throughout Europe?

Simple and complex refineries: patterns of location

To clarify the locational issue, refineries have been grouped on the basis of the complexity of their technological profiles in 1985. The first group, identified earlier as the modal group, comprises the 49 refineries operating two conversion technologies. For these the number in each country has been compared with the number expected from a constant proportional distribution based on each country's refinery population. The principal finding is that, with the exception of Italy and France, the deviation between the two figures does not exceed +/-2 in any country. With regard to the modal refinery type, therefore, the results do not suggest the existence of large-scale international contrasts.

While it is important to consider this typical group, however, it is arguable that the locational tendencies of simpler and more complex refineries may be of greater interest because these should be indicative of particularly high or low technological adoption levels. Simple and complex refineries accounted for 76 of the installations operating in 1985 (61 per cent). Fifty two fell into the simple category – those with no, or only one, downstream technology – leaving 24 refineries operating three or four technologies to form the complex group. Where simple refineries had conversion capacity, it was typically either vacuum distillation or thermal processing (30 cases). Only four supported either catalytic cracking or catalytic hydrocracking. Among the complex refineries the dominant technological combination was vacuum distillation, thermal operations and catalytic cracking (18 cases).

The location patterns of these groups (figure 6.3) reveal a number of contrasts. In France, extensive restructuring has meant that very few simple

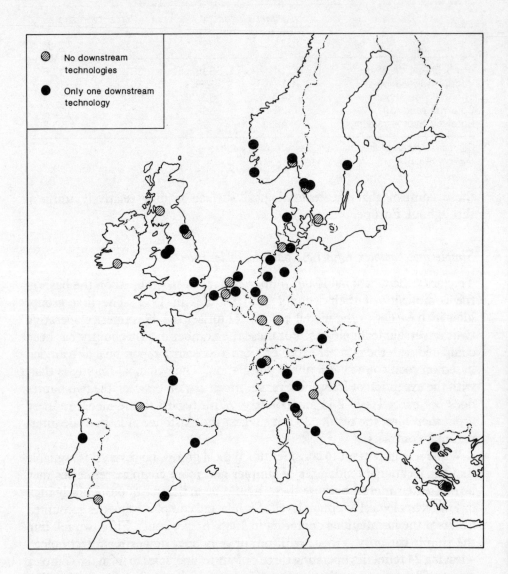

Figure 6.3(i)   Simple and complex refinery locations, 1985

refineries now remain. Conversely Norway, Sweden and Denmark possess ten simple refineries but only one complex one. Italy, meanwhile, exhibits intranational variations; refineries with a broad technological base are confined to Sicily and Sardinia, while simple ones are found only on the mainland. Beyond these differences, however, spatial biases are not obvious and

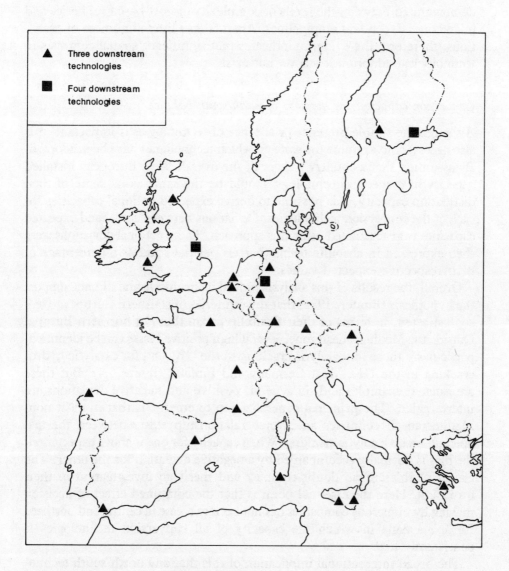

Figure 6.3(ii) Simple and complex refinery locations, 1985

certainly cannot be substantiated statistically. The two distributions show a considerable degree of overlap throughout most of Europe. In the majority of countries with a substantial number of refineries the proportions of simple and complex plants are close to the European averages. Similarly, techno-logical sophistication does not appear to be significantly associated with refinery site characteristics. For example, no significant difference can be demonstrated between the levels of complexity of port-based refineries and inland ones dependent on pipelines. Apart from a limited number of exceptions, therefore, the evidence indicates that significant spatial contrasts in technological adoption levels do not exist.

Conversion capacity: the question of north/south balance

Although the simple presence or absence of technologies is important, it is also necessary to examine the scale on which technologies have been adopted. By assuming that a country's share of the overall West European installed capacity of a specific technology should be the same as its share of total distillation capacity, it is possible to derive expected national capacities for each of the conversion technologies. Deviations between actual and expected capacities were calculated using this approach. These national deviations were then expressed in absolute terms (barrels per day) and as a percentage of their respective expected values.

Overall, the results of this analysis proved inconclusive and did not support the hypothesis (Pinder, 1986) that the refineries of southern Europe are less sophisticated, in terms of their flexibility, than those of northern Europe. Outside the Mediterranean zone, several high positive biases can be identified, particularly those for catalytic cracking in the UK, and for catalytic hydro-cracking in the UK, West Germany and Finland (figure 6.4). But these are isolated examples and, in general, positive and negative deviations are intermingled. The principal generalization to emerge is that in most non-Mediterranean countries an inverse relationship exists between thermal operations and catalytic cracking. When capacity for one is above expectation, the bias is normally accompanied by a negative deviation for the other. The causes of this are no doubt complex and merit an investigation in their own right. Here the essential point is that the combined effect of decision making by numerous companies has not been to create in central and northern Europe a zone in which the capacity of all conversion technologies is consistently high.

The broad international implication of this that any north/south technological contrast in Europe may be more apparent than real, is given further support when absolute, rather than percentage, data for southern Europe are considered. In table 6.5 the number of typical process units required to raise southern capacity levels to the European norms is calculated. In a few cases particularly vacuum distillation in Spain, the construction of several units would be needed. But in most instances a single average-sized unit would be sufficient to eliminate or erode substantially the percentage deviations

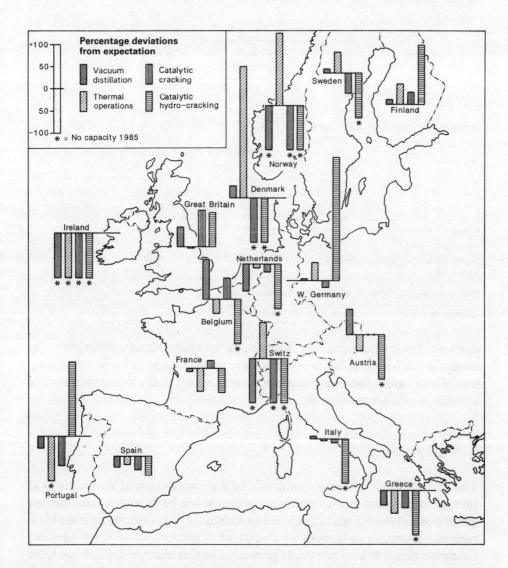

Figure 6.4 Conversion technology capacities: percentage deviations from expectation, 1985

in figure 6.4. In addition it must be noted that Spain and Portugal in contrast to most non-Mediterranean countries, have catalytic hydrocracking facilities. These facilities are admittedly limited: in both countries their conversion power is equivalent to only half the output of an average European catalytic

Table 6.5   Assessment of under-investment in southern Europe

	Datum calculations: European average unit capacities		
	Vacuum distillation	Thermal operations	Catalytic cracking
Total capacity (000 bpd)	3470	1629	1540
Refineries operating technology	82	59	55
Average unit capacity (000 bpd)	42.3	27.6	28.0
	Southern European deviations expressed as a number of average units		
	Vacuum distillation	Thermal operations	Catalytic cracking
Spain (10)	− 4.5	− 1.0	− 1.6
Portugal (3)	− 0.9	− 1.1	− 0.7
Greece (4)	− 1.6	− 0.8	− 0.6
Italy (23)	+ 2.4	− 0.5	− 1.0

Refinery populations are shown in brackets

Source: SUORD

cracker. Yet in Portugal this is sufficient to offset almost completely the catalytic cracking deficit, while in Spain the deficit is reduced to a single unit. Once again, therefore, the evidence does not point to the existence of a major north/south contrast.

A CRISIS RESOLVED?

The European Commission's attitude to the contraction of the industry is that, at least within the Community, progress has been smooth and has not generated a need for spatial policies to safeguard the interests of individual regions, countries or groups of countries (Commission of the European Communities, 1983). The results presented in this chapter indicate that, for Europe as a whole, this is also true of the investment wave. Although it is frequently argued that spatial technological lags may have important regional economic repercussions (Thwaites, 1982) variations in oil refining's technological surface do not appear sufficient to justify intervention by governments or supranational bodies. Even in the case of Norway, Sweden and Denmark, where the *number* of relatively simple refineries is high, the total *capacity* of existing conversion technologies is not excessively low. Similarly, although the Italian mainland lacks very complex refineries, it has many which operate two technologies and therefore compensate for the presence of simple installations noted earlier. In the broad context of the geography of technological change, therefore, European refining conforms with findings elsewhere that technological adoption rates are not heavily dependent on spatial factors (Gibbs and Edwards, 1985; Rees, Briggs and Hicks, 1985).

Yet this conclusion should not be allowed to obscure the fact that it is necessary to distinguish between the uptake of technologies and the achievement of goals to which the technologies are intended to lead. The one does not guarantee the other, although there is often an implicit assumption that this is the case. In making this distinction Vielvoye (1985) painted a picture of refining that was far from reassuring. Because of a continuing shift of demand away from heavy products, conversion capacity needed to rise still further: by 18 per cent between 1984 and 1990, and by another 12 per cent by the year 2000. Moreover, this would be necessary even though the current relationship between investment costs and product prices offered little prospect that future investments would show returns matching those of the highly profitable conversion units commissioned in the late 1970s.

By mid-1985 therefore, it seemed that expensive emulation strategies, though relatively easy for this major industry to adopt, were not capable of bringing a swift solution to the problem of harmonizing its product and demand profiles. The target had moved not suddenly, but on a long-term trajectory requiring equally long-term adjustment strategies. In mixed-scanning terms, after the reorientation forced by the 1970s oil price crises, incrementalism now appeared necessary to maintain momentum towards the new long-term goals. In the broad context of mature industry development, the indication was that satisfactory technological adaptation should be seen not as episodic but as a constant process of monitoring and adjustment.

Early in 1986, however, it became evident that refining's economic environment was once more evolving rapidly. Crude oil prices fell from $30 a barrel in November 1985 to $17 in February 1986, by which time the real cost of crude was at a level first reached mid-way through the 1973/4 oil price crisis. Predictions were made that prices would continue to slide, and the new trend was quickly labelled the third oil price crisis. Against this background one final question must be posed: will European refining be forced into a further episode of strategic reappraisal, or will the new directions be maintained?

The response must be speculative, but it is evident that the relationship between product prices and crude oil prices will be critical. If product prices were to decline even more rapidly than crude prices, a new refining crisis would be highly likely. Accelerated contraction and reappraisal of marginal or loss-making conversion units would both be strong possibilities. Yet it must be recalled that the recent decline in oil prices was initiated by surplus production rather than by a dramatic collapse of product demand. Although consumers will expect reductions in prices, therefore, demand may prevent a product-price slump. Indeed, it may be argued that modest price reductions may lead to an increase in demand through either fuel substitution or higher economic growth rates. These two possibilities may, however, have very different effects on refinery investment. If fuel oil once again becomes competitive with coal, major consumers such as electricity producers may switch from one source to another. As was shown during the 1984/5 UK miners' strike, this might well have a significant impact on the market for fuel oil, changing the demand profile so that the incentive for further investment in conversion diminished (Greif and Caarten, 1986). Alternatively, if lower prices were accompanied by generally high economic growth rates which left the profile unchanged, refining economics might well improve and further investment could be financially justified. Certainly, the industry will do its utmost to resist entry into an uncontrolled downward product-price spiral. Years of losses, cross-subsidization and investment in costly technologies have prepared companies to maximize on any improvement. It remains to be seen whether this can be achieved since, for the third time since 1973, the future of the industry was shrouded in uncertainty.

NOTES

1 The definition of Western Europe is that used by the OECD: Norway, Sweden, Finland, Denmark, West Germany, the UK, Austria, Switzerland, Italy, Spain, Portugal and Greece.
2 Although inflexibility became a major problem, it should not be confused with total rigidity. In the mid-1970s the production profiles of refineries with advanced conversion technologies were, of course, more closely geared to demand than the industry average. Also, output could in some instances be adjusted by the choice of crude oil. For example, Arabian heavy crude yields approximately 55 per cent residual fuel oil while others, such as Arabian light and most North Sea oils, yield 45 per cent or less.
3 The exact proportion will depend on refinery design and the crude oil that is refined. The proportions in this section are based on a crude yielding *c*.45 to 48 per cent residual fuel oil.
4 The effect of the catalyst is to accumulate carbon atoms from the vaporized hydrocarbon molecules. The carbon is burned off in the regenerator at 600°C before the catalyst returns to the reactor. This high energy input is then turned to advantage as the heated catalyst assists feedstock vaporization.

5 The situation in Norway, Sweden and Denmark is ameliorated by the fact that all but three refineries have at least one downstream technology. As is indicated at the end of this chapter, the result is that total conversion capacity in these countries is not unusually low. Despite this, compared with the rest of Europe the number of refineries with no more than one conversion technology can be shown to be significantly high, while in France the number is significantly low. (Chi squared = 11.6 degrees of freedom = 2; $P<0.005$).

REFERENCES

Bachetta, M. (1978) The crisis in oil refining in the European Community, *Journal of Common Market Studies* 17, 87–119.

Bergren, H. (1984) Shell develops new method to produce more fuels from intractible residues, *Rotterdam-Europoort-Delta* 4, 8–9.

British Petroleum (Annually) *Statistical Review of the World Oil Industry*, British Petroleum, London.

Commission of the European Communities (1983) *The Oil Refining Industry of the Community* COM Document (83) 304 Final, Commission of the European Communities, Brussels.

Commission of the European Communities (1985) *The Situation in the Oil Refining Industry and the Impact of Petroleum Imports from Third Countries* COM Document (85) 32 Final, Commission of the European Communities, Brussels.

Etzioni, A. (1973) Mixed-scanning: a third approach to decision-making. In Faludi, A. (Ed.) *A Reader in Planning Theory*, Pergamon, 217–30.

Gibbs, D. C. and Edwards, A. (1985) The diffusion of new production innovations in British industry. In Thwaites, A. T. and Oakey, R. P. (Eds.), *The Regional Economic Impact of Technological Change*, Frances Pinter, London, 132–63.

Grief, B. and Caarten, M. B. (1986) De derde oliecrisis, *NRC Handelsblad Weekeditie*, 4.2.86, 9.

Haines, B. A. (1978) *Petroleum Chemistry Refining Processes and Chemical Manufacture in Esso*, Esso Petroleum, London.

Le Heron, R. B. (1980) The diversified corporation and development policy: New Zealand's experience, *Regional Studies* 14, 201–17.

Malecki, R. J. (1980) Corporate organisation of R. and D. and the location of technological activities, *Regional Studies* 14, 219–34.

Molle, W. and Wever, E. (1984a) *Oil Refineries and Petrochemical Industries in Western Europe: Buoyant Past, Uncertain Future*, Gower, Aldershot.

Molle, W. and Wever, E. (1984b) Oil refineries and petrochemical industries in Europe, *GeoJournal* 9, 421–30.

Odell P. R. (1983) *Oil and World Power*, 7th edition, Penguin, Harmondsworth.

Pinder, D. A. (1984) Western European oil refining, economic crisis and locational change: some hypotheses tested, *Southampton University, Department of Geography Discussion Papers*, 27.

Pinder, D. A. (1986) Crisis and survival in western European oil refining, *Journal of Geography* 85, 12–20.

Plummer, D. P. (1973) The petroleum refinery. In Hobson, G. D. and Pohl, W. (Eds.), *Modern Petroleum Technology* Applied Science Publishers, London.

Rees, J., Briggs, R. and Hicks, D. (1985) New Technology in the United States' machinery industry: trends and implications In Thwaites, A. T. and Oakey, R. P. (Eds.), *The Regional Economic Impact of Technological Change*, Frances Pinter, London, 164–94.

Royal Dutch Shell (1983) *The Petroleum Handbook*, Elsevier, Amsterdam.

SUORD, Southampton University Oil Refinery Database. This comprises annual data on individual refineries published by leading industry journals such as *Oil and Gas Journal* and *Petroleum Economist*.

Thomas, M. D. (1985) Regional economic development and the role of innovation and technological change. In Thwaites, A. T. and Oakey, R. P. (Eds.), *The Regional Economic Impact of Technological Change*, Frances Pinter, London, 13–35.

Thwaites, A. T. (1982) Some evidence of regional variations in the introduction and diffusion of industrial products and processes within British manufacturing, *Regional Studies* 16, 371–81.

Vielvoye, R. (1985) Future gloomy for refining industry in Western Europe, *Oil and Gas Journal* 83, 11, 41–6.

7

Technical change and the restructuring of the North American automobile industry

J. Holmes

The objective of this chapter is to analyse some of the forms taken by recent restructuring and reorganization in the 'Canadian' automotive products industry.[1] The analysis focuses on one of the most striking features of restructuring in this established industry; the significant changes that are occurring in production technology and their consequences for the organization and locational structure of the industry. In the late 1960s and early 1970s the North American automobile industry could be characterized as being at a 'mature stage of development' with relatively stable product and process technologies. However, the industry subsequently experienced rapid technological change, especially after 1979, with respect to both product and process. The latter involves a combination of closely interrelated changes in both the technical and social organization of the production process which, in turn, appear to have significant implications for the future geography of the automobile industry both at an international scale and within North America.

There are numerous interpretations of the causes of the ongoing crisis in the North American auto industry (see National Academy of Sciences, 1982, pp. 13–16). This chapter argues that the crisis in the automotive sector represents just one specific instance of the more general restructuring crisis which confronts many sectors of industrial production in virtually all of the countries of the OECD. Thus, although the chapter is concerned with one specific industrial sector, it attempts to situate the forms taken by restructuring and technical change in the auto industry within the context of this more general restructuring crisis.

Clearly, the scale and nature of this restructuring is qualitatively quite

different from the processes of restructuring associated with the internal
dynamics of relatively stable periods of accumulation, such as the operation
of the 'normal' business cycles. With specific reference to the auto industry
a recent government report noted:

> the current crisis in the North American automotive industry is much
> more than a temporary downturn in an industry noted for its cyclical
> nature. . .at the heart of the current crisis lie fundamental changes in
> the nature of world markets and in the basis of competition for those
> markets (Canada, Federal Task Force, 1983, p. 45).

Restructuring on this scale is precipitated by major economic crises and occurs
in an attempt to create new conditions under which sustained profitable
accumulation will again be possible. Thus,

> [crises] compel capital to reorganize itself to prepare for new rounds
> of capital accumulation. . .capital accumulates through crises which
> become the cauldrons in which capital qualitatively reorganizes itself
> for future economic expansion (O'Connor, 1982, p. 312).

Such crises become a 'veritable hot house' for the development of the forces
and relations of production and often result in profound changes in the
organization of labour processes, labour markets and the industrial organization of production systems, as the search for new production technologies
and the conflict between workers and managers intensifies (Morgan and
Sayer, 1984, p. 4).

To understand the general structure and context of the economic relations
within which recent technical change in the North American auto industry has taken place, it is necessary first to understand the nature and
characteristics of the crisis which triggered the current period of restructuring.
Therefore, this chapter, which is divided into three main sections, begins
by briefly describing how this crisis resulted from the disintegration of
Fordism – the particular phase/model of capitalist accumulation which
had its origins in the 1930s and reached its zenith in the 1950s and 1960s.
The second and third sections of the chapter describe and contrast the
methods of production which characterized production in the North American
auto industry during the 'golden age' of Fordism with what appears to
be now emerging from the crisis as a new model of production technology
in the industry, stressing the changes in process technology which are
occurring.

ORIGINS AND NATURE OF THE CURRENT RESTRUCTURING
CRISIS IN THE NORTH AMERICAN AUTO INDUSTRY

There is mounting evidence that in the late 1960s the world economy entered the third 'long wave' crisis of the past hundred years and that the restructuring that is now underway is an attempt to find a 'solution' which, if found, will herald the emergence of a new and qualitatively distinct phase of capitalist accumulation (Mandel, 1980; Aglietta, 1979; Armstrong et al., 1984). This chapter adopts the interpretation developed by the French 'regulation school' of the formation and eventual collapse of the economic conditions which fostered the post-1945 boom (Aglietta, 1979; De Vroey, 1984; Lipietz, 1985; 1986). Central to the regulation school's analysis is the development and subsequent crisis of a model of capital accumulation which they label the regime of intensive accumulation, or 'Fordism' for short.

The Crisis of Fordism

In the first two decades of the twentieth century the application of the principles of work study and scientific management, developed by Taylor, and of the assembly-line system for organizing industrial production, which was adopted and further refined by Ford, laid the ground work for a dramatic and sustained rise in labour productivity in the sectors producing consumer goods. The emphasis was on the production of goods capable of being produced in large volumes which would yield considerable scale economies and a progressive lowering in the unit costs of production. However, the full development of intensive accumulation based on Taylorism and Fordist techniques of production organization was blocked in the first half of the period between the two World Wars by a lack of adequate purchasing power on the part of the working class, and this led to the massive and unprecedented crisis of overproduction in the late 1920s.

The massive restructuring which occurred in the 1930s and early 1940 in the wake of that crisis ushered in a new mode of regulation by radically transforming the spheres of consumption and reproduction. To be successful, mass production based on Fordist production methods required as a corollary the development of mass consumption. This was eventually achieved in the post 1945 period with the full development of Fordism, but its roots can be traced to a number of fundamental and crucial changes in the norms and patterns of consumption which were set in motion in the mid-1930s, in part to supply outlets for new types of consumer goods and thus regulate and harmonize the production of consumer and capital goods, but also to enable the satisfactory reproduction of workers suited to the new forms of work associated with Fordism.

The progressive extension and generalization of Fordist techniques of production and the continued expansion of mass consumption led to steady economic growth and rising levels of productivity in the period from 1945 through to the late 1960s. Two features of Fordism should be emphasized: first, the crucial significance of rising rates of productivity and the link between the latter and the growth in real wages, second, the fact that Fordism developed as a regime of accumulation whose regulation was essentially *internal* to each national economy. It was predicated upon the internal transformation of industrial production processes and on the growth of the internal domestic market via the development of mass consumption and rising real wages linked to productivity growth.

The onset of the current economic crisis in the late 1960s and early 1970s was the result of the breakdown of a number of the key regulatory mechanisms which had been instrumental in the successful rise of Fordism. In particular, there was a dramatic slowdown in productivity growth and a sharp intensification of competition at the international level.

The slowdown in productivity growth is particularly crucial to our analysis, since it has been the need to restore a sustained growth in labour productivity which has made technical change in industries such as the auto industry so imperative. In Canada, the rate of increase in labour productivity (measured in terms of output per person hour) in the manufacturing sector as a whole slowed dramatically in the period from 1973–82, after showing strong growth throughout the period from the late 1940s, to the early 1970s, (see table 7.1).

Table 7.1 Annual percentage changes of labour productivity and unit labour costs in Canadian manufacturing, 1961–1984

Period	Output	Person hours	Labour Compensation	Output per person hour	Unit labour cost
1961–1964	3.9	0.6	10.1	3.3	5.9
1961–1968	7.3	2.8	8.4	4.4	1.0
1968–1973	6.0	0.9	9.1	5.0	2.9
1973–1978	2.2	−0.1	12.5	2.1	10.3
1978–1982	−1.8	−2.0	9.8	0.1	12.0
1982–1983	6.0	−0.4	6.3	6.4	0.3
1983–1984	8.3	4.2	5.8	4.0	−2.3

Source: Statistics Canada: Aggregate Productivity Measures, cat. 14–201
        Statistics Canada Daily, March 25, 1985

At a disaggregated level, the data for the auto industry exhibit a similar pattern (see table 7.2). One consequence of the slowdown in productivity growth was a rapid increase in unit labour costs after 1973 (see table 7.1). A number of factors have been suggested as contributing to this dramatic slowdown in productivity growth (Holmes and Leys, 1986). However, irrespective of what led to the slowdown, it resulted in a crisis of profitability as the labour cost per unit output continued to rise and as further increases in the technical composition of capital failed to be offset by corresponding increases in productivity (Zohar, 1982, p. 31).

Table 7.2 Labour productivity, Canadian automotive products industry, average annual rate of increase, 1956–1983

Period	Average annual percentage increase in real output/production workers hour	
	Motor Vehicle Assembly Sector (SIC 323)	Motor Vehicle Parts Sector (SIC 325)
1956–1983	5.0	4.2
1956–1961	7.0	2.0
1961–1966	4.2	4.5
1966–1971	7.5	7.0
1971–1976	1.5	4.0
1976–1981	1.5	−1.7
1981–1983	13.1	17.0

Sources: Statistics Canada Catalogue 42–209 Various years
Statistics Canada Catalogue 42–210 Various years
Statistics Canada Catalogue 62–453 Industry Selling Price Indices 1956–76
Statistics Canada Catalogue 62–011 Industry Selling Price Indices. Various years

The intensification of international competition resulted from the very success of Fordism in the 1950s and 1960s and, in particular, from the inherent tendency within Fordism towards overcapacity. The competitive pressure to increase the scale of production within individual plants coupled with a levelling off in demand as markets became saturated led, by the mid-1970s, to considerable overcapacity in such sectors as autos, steel and textiles. The competition among the major capitalist economies for each other's markets became intense, since it was precisely in those markets that levels of mass consumption had been transformed most profoundly by Fordism. The intensification of competition and the existence of substantial variations in unit costs of production between different national producers not only led to an increased downward pressure on rates of profit but, in some manufacturing sectors, threatened the very existence of future production in countries such as the US and Canada. The latter had the higher unit costs of production and little hope of reducing them by increasing labour productivity with *existing* process technology.

One response to this crisis has been for industry to instigate a large scale programme of restructuring and reorganization involving two general strategies. Initially there was a tendency for firms in certain industries further to internationalize their production by moving production 'offshore' to lower cost sites in the Newly Industrializing Countries (NICs) of south east Asia and Latin America, and particularly to the free trade zones located in those regions. The consumer goods produced in the NICs were mainly for export to the advanced capitalist countries and contributed to the development of what has been labelled the new international division of labour (NIDL) (Frobel et. al., 1979). However, as Lipietz (1982) has pointed out, whilst the strategy of geographically relocating production may have served in the short run to increase the average rate of surplus value for particular firms by reducing the cost of variable capital, it neither brought about a significant expansion in markets nor served to increase productivity. Consequently, it did not basically resolve the crisis of Fordism and the problem of how to give a new impetus to productivity and to consumption on a world-wide scale.

The second response has involved rapid changes in process technology with the aim of responding to changing competitive conditions and, in particular, to increasing labour productivity and reducing unit costs. This tendency has been heightened by both the limited solution offered by the further internationalization of production and by the continued escalation of unit labour costs in North America right through to the early 1980s.

The development of robotics and other microprocessor-based machine tools and process control systems has created the potential for a radical restructuring of the labour process and the achievement of significantly higher levels of potential and actual labour productivity. Perhaps the most important feature of robotics and computer numerically controlled (CNC) machine tools is their *flexibility*, one consequence of which is that *the link between lowering unit product costs and long production runs based on economies of scale has been considerably weakened, if not completely severed*. This clearly has far reaching consequences for both the organizational structure of manufacturing and for location theory.

In addition, but closely related, to these changes in 'hard' process technology (such as faster machining, automatic welding and robots) there have also been changes in so-called 'soft' technology, involving the broader technical and social organization and management of the manufacturing process. In fact, Fraser (1982, p. 4.) argues that in the auto industry '...changes in soft technology have been and will be more important than changes in hard technology.'

The preliminary data on output and employment in Canadian manufacturing for 1983 and 1984 (tables 7.1 and 7.2) appear to provide the first firm evidence of the qualitative transformation of production that is under way. They show sharp increases in labour productivity and declines in the rate of growth in unit labour costs (note that 1984 saw the first absolute decline in unit labour costs in Canadian manufacturing since 1964). The evidence for the US shows a similar trend (Fulco, 1984).

Given our view that the economic slowdown has profound social and political roots, it is important to emphasize that the resolution of the restructuring crisis requires not only changes in production technology but also broader social and political change (Evans and Kaplinsky, 1985, p. 6). Therefore, equally important and clearly interlinked with the process of industrial restructuring and technical change is the need to transform virtually every aspect of what the French regulation school refer to as a 'regime of accumulation', i.e. patterns and norms of consumption, labour relations, the labour market, wage determination, the monetary system etc. Davis (1984, 1985) provides interesting evidence that supports the view that such far reaching changes are presently underway in the US.

The Crisis of Fordism in the Canadian Automotive Products Industry

There are a number of reasons why the automotive products sector is particularly pertinent to a study of recent technical and industrial change in Canada. First, it occupies a position of prime importance within the industrial structure of Canada. It is the largest manufacturing sector employing over 110 000 people directly and, over the last ten years, automotive exports have contributed about 20 per cent of Canada's total merchandise exports and almost 60 per cent of Canadian exports of manufactured end products. Second, the very term Fordism underscores the fact that historically the automotive sector has been a 'leading' sector in terms of the development of new process technologies and forms of labour process organization. Just as it was important in the development and introduction of mass production transfer line technology earlier this century, so today the auto industry is one of the leading sectors in the application of microelectronic technology to production and process engineering. Finally, the development of the Canadian automotive sector in the period after the Second World

War demonstrates both the nature and limits to Fordism as a mode of regulation.

Figure 7.1 shows, for the period 1955–83, the general trends in output, employment and productivity for the Canadian auto assembly and automotive parts sectors. These data reveal that there have been, in fact, two crises and associated periods of restructuring and reorganization within the Canadian automotive products industry during the last thirty years. The first was a *sectoral* crisis in accumulation, confined to the Canadian automotive products industry, which developed in the late 1950s and early 1960s and interrupted the long post-1940s boom. This crisis resulted from the full development of Fordism in the Canadian automotive sector being blocked by limited market size (Holmes, 1986a). It was resolved by the negotiation in 1965 of a conditional free trade agreement – the Auto Pact – betwen Canada and the US. This enabled the reorganization and rationalization of automotive production in Canada and its integration with production in the US to form one integrated production system in which production and consumption were 'regulated' at the level of the combined US and Canadian economies (i.e. the 'continentalization' of Fordism). Subsequently, the period 1962–70 saw renewed expansion and growth in the Canadian automotive industry and the development of a clear spatial division of labour within the North American auto industry in which Canadian plants were 'allocated' those parts of the production process which had a high labour content (see Holmes, 1983).

The second crisis, which affected not just the Canadian portion but the whole of the North American automotive industry in the late 1970s, is simply one manifestation of the *general* crisis of Fordism described at the beginning of the chapter. Although the extent of the downturn which occurred in 1979–82 was remarkable – output fell by over 30 per cent and employment by over 20 per cent (figure 7.1), it was simply the most pronounced moment in a decade of increasing economic uncertainty and deepening crisis, the origins of which can be traced back to at least the early 1970s. Nevertheless, when viewed in the context of the world auto industry as a whole it is important to distinguish between the early-mid 1970s and the late 70s-early 80s. In the first period, although there had been a dramatic slowdown in the rate of growth of labour productivity in the North American auto industry and there were sharp cyclical fluctuations in employment and output within national auto industries, including those of Canada and the US, employment levels and market shares between the major auto-producing countries remained remarkably stable. In sharp contrast, the late 1970s-early 1980s saw the end of this stability as international competition intensified and North American auto markers faced serious competitive challenges from offshore producers, and particularly from those in the Far East (Altshuler, 1984, p. 200).

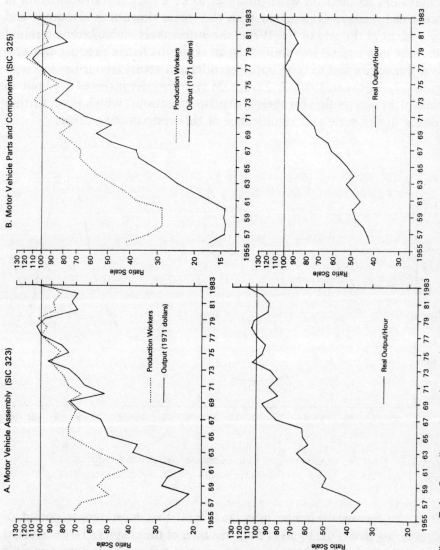

Figure 7.1 Canadian automotive products industry, employment, output and productivity, 1955–1983
(1977 = 100)

The initial response of the North American auto industry to this crisis involved the further internationalization of production, but, more recently, attention has focused on what promises to be a radical transformation in process technology. Concurrent with the major downturn in output and employment at the end of the 1970s, the automakers embarked on a round of massive new capital investments in an effort to change product and process technologies and to raise both potential and actual labour productivity (figure 7.2). Between 1980–2, Ford, GM and Chrysler invested $3.2 billion in new plant and tooling for their Canadian operations, which is yet further evidence of the scale and significance of this restructuring crisis.

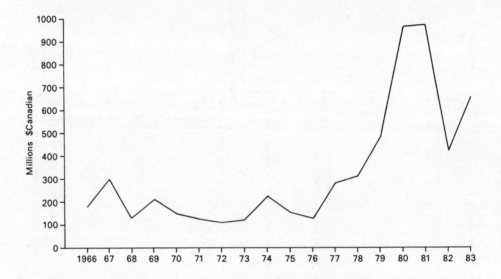

Figure 7.2 New capital expenditure in the Canadian automotive industry, 1966–1983
Source: Statistics Canada: Capital and Repair expenditures, manufacturing subindustries, Cat. 61–214.

It is the technical change associated with this most recent period of restructuring that is the focus of the remainder of the chapter. To understand the extent and nature of these changes in process technology and their possible consequences for the future organization and geographical structure of the auto industry in Canada, the two remaining major sections of the chapter contrast the organization of production which characterized the North American auto industry during the 'golden age' of Fordism in the 1950s and 1960s with what appears to be emerging in the mid-1980s as a new model of production technology in the industry. It is important to stress the strong inter-relationships that exist in both 'models of production' between the nature of the product market, the competitive context within which auto-makers operate, the nature of the labour process, in both a technical and social sense, and the broader industrial organization of production in the industry. The contrasts in these characteristics between the two models are shown schematically in figure 7.3 and discussed in detail in the following two sections.

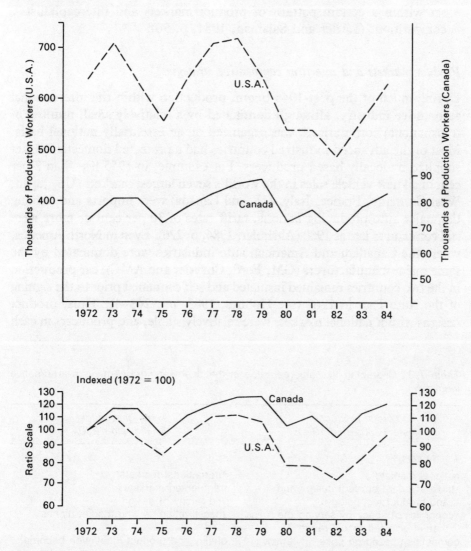

Figure 7.3   Auto-industry production workers, Canada and USA, 1972–1984
Source: Statistics Canada: US Bureau of Labor Statistics, Employment, earnings and hours Cat. 72-002

CHARACTERISTICS OF PRODUCTION IN THE NORTH AMERICAN
AUTOMOBILE INDUSTRY DURING THE
'GOLDEN AGE' OF FORDISM

> Taylorism and Fordism as forms of work organization are constrained
> by certain economic and technical limits and...in particular they are
> set within a certain pattern of product markets and intercapitalist
> competition. (Littler and Salaman, 1984, p. 90).

Product markets and interfirm competitive strategies

During much of the post-1945 boom, production within the international
automotive industry, although dominated by a relatively small number of
transnational corporations, was organized on an essentially national basis.
Most of the advanced industrial countries had a protected domestic market
supplied by locally based producers. For example, in 1955 less than 2 per
cent of all new vehicle sales in the world's seven largest markets (US, Japan,
West Germany, France, Italy, UK and Canada) were imports and, among
the major auto producing nations, tariff rates of 30 per cent or more were
still common as late as 1960 (Altshuler, 1984, p. 220). Even in North America,
where the Canadian and American auto industries were dominated by the
same major manufacturers (GM, Ford, Chrysler and AMC), car production
in the two countries remained insulated and self contained prior to the signing
of the Auto Pact in 1965 (see Holmes, 1983, pp. 263–4). Thus, product
designs within national markets were relatively stable, and producers in each

Table 7.3 Characteristics of production in the post-war north American automotive
industry

Fordist production (late 1930s–mid 1970s)	Emerging model in the mid 1980s
1 Markets	
national markets	international markets
stable standard product design and and product technology	fragmented markets diversity of design
economies of scope by serving the mass market with full model range	specialization serving market niches
competition based on price, style and and marketing	quality and product innovation become important competitive factors alongside price

Table 7.3   continued

Fordist production *(late 1930s–mid 1970s)*	*Emerging model* *in the mid 1980s*
2 Technical organization of the production process	
dedicated specialized machinery	flexible 'general' machines robotics and CNC tools
large scale scrapping of fixed capital with each model change	reprogrammable machines, capital costs initially high but capital saving in longer run
unskilled repetitive tasks performed by a 'deskilled' labour force	more varied tasks, a 'reconstitution' of skill
substantial production economies	production economies small
long production runs needed to achieve efficient production, rapid rise in MES	medium sized volumes of different models can be produced efficiently
scale economies in standardized components leading to specialization and rationalization between plants	components can be efficiently produced in a number of smaller plants
3 Social organization of the production process	
inflexible and hierarchical division of labour within the plant	flexible workforce with reduced levels of supervision
quality control through strict supervision	individual workers responsible for quality and improving production methods – use of quality circles
large buffer inventory stocks of components	just-in-time delivery of components
'fordist' labour contrct incl. annual increases in base wage rates, multiple job categories single level wage structure	wage increases linked to performance and wage rates to seniority, fewer and more general job categories
price oriented 'arm's length' relationship with component suppliers	closer association with suppliers – single sourcing, long term contracts, R&D
4 Industry structure	
large scale rationalized plants	smaller scale production possible
increased concentration and vertical integration	survival of medium sized producers possible and vertical disintegration
fewer producers	more producers in long run but with joint ventures in technology and production
5 Location and geography of the industry	
more geographically extensive production system – the 'World Car' scenario	tendency for reconcentration of suppliers around assembly points – 'regional' clusters of car production
location of production and trade determined by factor costs	location of production and trade determined by technological factors
possible shift to NICs to gain access to protected markets or for lower wages	automobile production remains concentrated in the OECD owing to technical advantage

country concentrated on designs and sizes suited to their primary national markets. Whilst the international trade in automobiles was growing it was still relatively small and most imports were confined to market segments that domestic producers could afford to ignore (Altshuler, 1984, p. 4). For example, right through to the late 1970s cars produced for the mass North American market were essentially a different product to those produced by European and Japanese car makers, and the response of North American producers to import competition was to differentiate still further their product range from those cars produced elsewhere in the world and to concede the bottom end of the market (i.e. small cars) to foreign producers.

Product technology, particularly with respect to major components such as engines, transmissions and suspension systems also remained stable within the North American auto industry with the only significant year-to-year changes occurring in body styling. One consequence of these stable product technologies was that competitive strategies between automakers within the Canadian and US markets focused on the marketing of differences in body styling and, perhaps most importantly, on price competition. Each major manufacturer aimed for economies of scope by serving the mass market with a full model range, and for each model there was overwhelming competitive pressure to lower unit labour costs through the achievement of economies of scale.

Technical organization of the production process

This competitive strategy had direct implications for the organization of the labour process. Because of standardized and stable product technologies, production systems were able to use higher and higher levels of mechanization for the production of components such as engines and transmissions. The competitive pressure to lower unit labour costs coupled with the trend towards greater integration of the production process had

> the effect of emphasizing the importance of scale of production as the ratio of constant to variable capital rose. A change in model not only became more expensive but involved extensive retooling and at least initially lower productivity until one again moved down the learning curve. (Jones, 1985, p. 9)

This underscores the classic dilemma which Abernathy (1978) identified in his pathbreaking study of innovation in the auto industry, between, on the one hand, the need to engage in product innovation in order to remain competitive and the need for higher productivity on the other, also to remain competitive. Abernathy argued that because of the specific characteristics of the North American mass market, North American producers were able to concentrate on the latter strategy at the expense of the former, but that

> production processes designed for efficiency, offer higher levels of productivity but they also become mechanistic, rigid, less reliant on skilled workers and more dependent on elaborate and specialized equipment. (Abernathy, 1978, p. 4)

One consequence of the use of dedicated specialized machinery was that model changes usually entailed the large scale scrapping of capital goods.

Thus, throughout the 1950s and 1960s the use of Fordist production methods pushed minimum efficient operating scales (MESs) in the car industry ever higher (UNCTNC, 1983, p. 73). During this period

> most of the productivity growth in the automotive industry, which was substantial in that era, was due to the design and construction of ever more productive facilities with specialized tooling and automation to achieve maximum economies of scale. As each new plant came on stream, that particular firm's productivity would increase, but it was a temporary advantage that lasted only until companies built ever more productive facilities. A sustainable advantage in manufacturing costs could only come from achieving higher market penetration and hence greater production volumes. (Canada, Federal Task Force, 1983, p. 51)

This quotation clearly points to the contradictory nature of this process. The tendency toward ever increasing MESs coupled with the levelling off in the world demand for cars due to market saturation within the countries of the OECD and very slow growth in demand in the developing countries led to substantial overcapacity in the industry by the mid-1970s. In turn, this led to an intensification of competition among automakers at the international level, particularly for the markets of the OECD. Thus, in the 1970s the annual growth rate in demand for autos in the OECD fell to 1.6 per cent (compared to over 6 per cent in the previous decade) and the percentage of new vehicle sales in the seven largest national markets which were imports, increased from 2 per cent in 1953, to 14 per cent in 1970, and to 24 per cent in 1981. Between 1978–82 world demand for cars declined by 13 per cent, but world trade in automobiles increased by over 30 per cent (Canada, Department of Regional Industrial Expansion, 1984, p. 38).

One direct consequence of this necessity for firms to achieve economies of scale and higher productivity through increasing the length of production runs was a tendency towards increased specialization and rationalization of production between plants. Since different stages of the automobile production and assembly process have different minimum efficient scales, there were marked cost advantages to segmenting the production process and integrating production between geographically separate plants, so that each component and assembly plant could achieve the full benefits of economies of scale at their stage of the manufacturing process.

Other strategies which emerged in the 1970s as firms sought to eke out the maximum potential productivity from production methods which remained essentially Fordist in design included the following: a reduction in the number of separate parts used in the production of a particular vehicle; design coordination and joint ventures between companies with the aim of increasing the number of interchangeable parts between different models and even between models assembled by different companies; attempts to rationalize a firm's model ranges towards one basic model per size class. These strategies all aided in the attainment of longer production runs of particular parts, higher productivity and lower unit costs.

The technical organization of production just described carried with it clear implications for the present and future structure and geography of the auto industry under Fordism. There would be a tendency toward ever larger-scale rationalized plants, increased vertical integration, and continued concentration of ownership which would result over time in fewer and fewer producers. Production systems would become more geographically extensive and dispersed as the location of production and patterns of trade in the industry continued to be determined by the search for lower factor costs. In fact, the 'world car' concept represents Fordism taken to its logical conclusion within the world automotive industry – a single standardized design on a global scale to be produced at the cheapest possible cost (Friedman, 1983, p. 367).

Social organization of the production process

Another set of distinctive characteristics of standardized Fordist production in the North American auto industry concerned the social organization of production and included the technical and social division of labour on the shop floor, the formal system of labour relations and collective bargaining, and the wider relationships which existed between component suppliers and subcontractors and the major auto assemblers.

Standardized production systems resulted in a division of labour within the factory which was inflexible, hierarchical and characterized by increasing automation, routinization, and deskilling of production tasks (Gartman, 1982). The thorough application of Taylorist principles of scientific management broke down assembly line tasks into almost foolproof steps to be performed repetitively by unskilled line workers. The high degree of routinization virtually eliminated any opportunity for responsible worker involvement or initiative. Under this system

> it was assumed that the worker would not report on problems, would not repair his own machines and would take no initiative for spotting or correcting faults...high quality could only be maintained in such a situation by having an army of supervisors, checkers and repairers, while improvements in the production process were the responsibility of another set of specialists, the production engineers. (Jones, 1985, p. 139)

The problems associated with quality control under this sytem not only required additional layers of direct supervision on the shop floor but, because even with all the supervision it still resulted in a relatively large number of defective parts, it also had an important consequence for the technical organization of production. It meant that large inventories of parts had to be held as buffers between each major production operation, both within the final assembly plant and between the assembly plant and its suppliers, in order to ensure that parts of the production system could continue to operate whilst faults were diagnosed and corrected in other parts of the system. The holding of large buffer stocks between the assembly plant and component suppliers was also seen as protection against possible disruption of final assembly by labour problems in supplier plants. Another strategy employed by auto assemblers to safeguard the continuity of component supplies was the multisourcing of parts. The latter became a key feature of the industry in the late 1960s and early 1970s as the incidence of 'wildcat' strikes in the auto industry increased.

The pattern for post-1945 labour contracts in the North American auto industry was set by the 1948 contract between the UAW and GM (Katz, 1984; Stone, 1981). Under these contracts annual increases in base wage rates became virtually automatic, but in return production-related matters such as line speeds, work standards and the design of jobs remained the almost sole prerogative of management. The annual wage increases which were negotiated in three year national agreements consisted of two parts; a cost of living adjustment escalator (COLA) to protect real wages against price inflation, and an annual improvement factor (which after the mid 1960s was set at 3 per cent) linked explicitly to expected increases in productivity. This wage formula provided continuity in wage determination in the industry across time and across the industry at any given point in time and except for very minor adjustments the formula rigidly set wages among the 'big three' from 1948–1979 (Katz, 1984, p. 206). Two other important features of these contracts were a complex set of strictly demarcated job categories and the fact that whilst job security was tied to seniority, wage rates were not.

In the labour contracts, the annual improvement factor clause geared the annual increase to the growth rate of productivity for the economy as a whole rather than to productivity increases in the auto industry (Katz, 1984, p. 207). Throughout the 1950s and 1960s the productivity increases enjoyed by the auto industry were generally substantially above the average for the economy. Therefore, during this period, the annual increases in base wage rates in the industry were easily financed by increases in labour productivity. However the slowdown in labour productivity growth which occurred in the early 1970s, coupled with continued annual increases in wage rates as a result of the rigid application of the wage formula, was a major factor contributing to the rapid escalation of unit costs and associated decline in profitability. Whilst, like the competitive strategy to which it was linked, it contained the seeds of its own destruction, most writers agree that this system of labour-management relations:

> seems to have provided a workable approach to workforce management throughout the twenty year period from 1945 to 1965. At the plant level, the primary competitive objective was production, an objective well suited to methods of control based on machine pacing, supervision and work standards . . . as conditions and people changed in the late 1960s, it became increasingly unprofitable. In restrospect it is clear that while the traditional employment relationship gave the worker a limited voice in setting conditions of work and while it paid well it did not secure loyalty. Nor was it intended to . . . production objectives required people to operate the equipment, but as long as minimum standards were met (and supervisors ensured they were) loyalty or commitment were not essential. (National Academy of Sciences, 1982, p. 115)

Standardized production also encouraged assembly firms to develop price oriented 'arm's-length' relationships with suppliers in which only the price of the components and a reasonable assurance of uninterrupted supply were important. The frequent competitive bidding by suppliers for orders and the multisourcing of parts were used to increase competitive pressure on suppliers to lower unit costs and prevent production disruptions in the assembly plant.

These then were the primary features and characteristics of the relatively stable model of production which operated successfully in the North American auto industry from 1945 through to the early 1970s. The eventual demise of Fordism in the late 1970s was the result of a combination of factors. Certainly contingent factors, such as the rapid rise in gasoline prices through the 1970s, and the slump in demand associated with the adoption of mone- tarist policies in North America and Western Europe played a role. But, first and foremost, the collapse was caused by the coalescence and mutual reinforcement of those tendencies which we earlier identified as being not only central to Fordism but also contradictory; namely, the tendency towards overcapacity in the industry which led to intensified international competition, and the particular form of wage determination in the industry which led to rapidly escalating unit costs when productivity increases could no longer be sustained.

This situation raised the possibility of dramatic shifts in the location of employment within the international auto industry, since national auto indus- tries were very differentially positioned to deal with these new competitive conditions. The productivity and cost differentials that existed between the North American and Japanese automotive industries were enormous. Drawing upon reliable and oft quoted data produced by Abernathy, Perry (1982) summarized the competitive position of the Japanese automotive industry relative to North America in mid-1981 as follows:

> total hourly compensation in the Japanese automotive industry was estimated to be 62 per cent of that in Canada and 46 per cent of that in the US...in the period from 1978 to 1979 Japanese producers required an average of 80.3 man hours per vehicle while US producers required 144 man hours. (Perry, 1982, p. 31)

Abernathy estimated that, in 1979, the combination of lower hourly compensation and superior productivity of both direct and indirect labour yielded a landed cost advantage in the US market to the Japanese producers of $US 1650 per vehicle (quoted in Perry, 1982, p. 32). A similar calculation by Perry for Canada in 1981 yielded an estimated differential of $C 1882 per vehicle in Japan's favour (Perry, 1982, p. 33). Despite the difficulties involved in estimating the absolute productivity and cost differentials that exist between North American and Japanese auto producers (see National Academy of Sciences, 1982; Norsworthy and Malmquist, 1983; Cole and Yakushiji, 1984) there is agreement that the differentials are substantial and favour Japanese producers. Moreover, the threat of Japanese competition is not confined to the advanced industrial economies since

> the completely new standards of organizational efficiency established by the Japanese have also pulled the rug out from under the feet of the developing countries...as a result even the South Koreans, with a $1 an hour wage rate in 1980, cannot produce a comparable vehicle for the same costs as the Japanese with a $7 an hour wage rate. (Jones, 1985, p. 141)

'AFTER JAPAN': TOWARDS A NEW MODEL OF PRODUCTION IN THE NORTH AMERICAN AUTOMOBILE INDUSTRY

Earlier, two general tendencies which had emerged in response to the crisis of Fordism were noted, namely, the further internationalization of production, and, second, concerted efforts to raise productivity through the development of new 'hard' process technologies and new forms of labour process organization. Both tendencies are evident in the recent restructuring of the auto industry and each has significantly different, and to some degree contradictory, implications for the future geography of production in the automobile industry at both the world scale and within North America. For example, at the world scale, in order to gain access to certain protected national markets it might be necessary for automakers in return to agree to produce and export parts or assembled cars from these countries, thus leading to a more dispersed pattern of production in the industry. However, the potentially large in-transit inventories and quality control problems associated with this strategy run counter to the rapidly emerging best production practice in the auto industry which is leading to a geographical reconcentration of production.

Internationalization of production in the auto industry has taken the form of integration within and between the OECD countries and the development of worldwide sourcing of some parts and subassemblies (e.g. engines, transmissions and wheels) (Johnston, 1982; Dankbaar, 1984; Cohen, 1983). Given the relatively high capital intensity of the automotive industry, the wage differentials that exist between the lesser developed countries and the advanced industrial countries appear to be relatively unimportant (in comparison with other industries such as clothing and semiconductors). Such factors as gaining access to cheap sources of energy and raw materials to reduce the value of constant capital, gaining access to markets still partially protected by tariff and non-tariff barriers, or the availability of state subsidies to socialize the costs of fixed capital have been much more important in shaping the internationalization of production in the automotive industry (Jenkins, 1984; Ward, 1982; Hawkesworth, 1981).

Two features of this internationalization process are of particular significance for the future shape of the auto industry in Canada. First, given the relatively low cost of energy in Canada, there has been a significant investment in new plants producing parts and components which have a high energy content, for example, parts made from aluminum or plastics. Second, when viewed in the context of the differential in labour costs that exists in the auto industry between Canada and the US, the fact that the final assembly of cars is likely to remain highly concentrated within the existing auto producing regions of the OECD implies that Canada will remain an important site for the final assembly of cars and the production of components which have either a high labour or energy content (Holmes 1986b).

With the wholesale movement of production to offshore low wage areas precluded for the reasons just stated, North American producers have concentrated on the development of new process technologies. the main objectives of the latter have been the cutting of production costs by both increasing labour productivity and engaging in other cost cutting measures, improving quality control, and increasing the flexibility of production systems so that product differentiation and quality can become important competitive strategies alongside price.

Obviously the paramount factor that has strongly influenced both the scale and the nature of technical change in the North American car industry since 1977 has been the serious challenge mounted by Japanese producers to the dominance of the North American market by domestic producers. Increasingly North American auto makers have looked to Japan as the provider of ideas and productivity targets which must be met if the domestic North American auto industry is to survive in anything like its present form. Although Ford has been the most explicit in this respect with its 'After Japan' (AJ) campaign which it initiated in 1979, virtually every American and European auto producer is now moving to incorporate key elements of Japanese production methods into their operations.

Although there has been considerable debate among industry analysts concerning the 'source' of the Japanese productivity and production cost advantage, there now seems to be a consensus that the advantage lies primarily in the 'social organization and management of the production process' which results in higher process yields[2] and, hence, higher labour productivity, and, second, in lower wage rates but *not* as a result of greater capital investment or more sophisticated 'hard' process technologies (Altshuler, 1984; Canada, Federal Task Force, 1983; national Academy of Sciences, 1982). Thus, in the 1960s and 1970s Japanese manufacturers established a new standard of best practice for the world automotive industry by demonstrating that dramatic productivity growth and cost savings could be obtained in ways other than the relentless pursuit of economies of scale. Consequently, 'the social organization of the production process, long thought to be fully perfected and "rationalized" along classical Ford and Sloan lines is now in dramatic flux.' (Altshuler, 1984, p. 183)

Although particular elements of the Japanese production management system, such as the quality circle and kanban/Just-in-Time (JIT) delivery systems, are being widely promoted by industrial management in North America, there is still considerable misunderstanding concerning the overall nature and structure of the so-called Toyota production system.[3] The basic idea of the latter is to maintain a continuous flow of products in factories in order to adapt production flexibly to changes in demand. The realization of such production flow is called Just-in-Time production at Toyota and entails 'producing only necessary items in a necessary quantity at a necessary time' (Mondon, 1983, p. vi). The optimal operation of the Toyota production system depends upon the implementation of a number of methods related to batch size, quality control, production smoothing, reducing setup times and the development of a more 'flexible' workforce. Though the main purpose is to reduce costs '...the system also helps increase the turnover ratio of capital (i.e. total sales/total assets) and improves the total productivity of the company as a whole'. (Monden, 1983, p. 1). Monden is probably correct in his claim that '...it would probably not be overstating our case to say that this [the Toyota production system] is another revolutionary production management system. It follows the Taylor system (scientific management) and the Ford system (mass assembly line)' (p. 1). Certainly, there is little doubt that the whole direction of management thinking in North America (and Western Europe) has been fundamentally altered by what the Japanese achieved in the 1960s and 1970s.

Seldom are innovations in production technology diffused simply through imitation. Rather 'the diffusion of innovations more typically involves their rejection, their modification and their partial selection'. (Clark and Tann, 1986, p. 2). For example, a number of studies of the British auto industry in the inter-war and immediate post-war period have stressed the divergence between the form in which Fordist mass production technologies were first developed in North America and the form in which they were developed in Western Europe (Lewchuk, 1983; Littler, 1982; Clark and Tann, 1986). Monden suggests that the Toyota production system arose out of the 'market constraints which characterized the Japanese automobile industry in post war days: great diversity within small quantities of production. Toyota thought consistently from about 1950 that it would be dangerous to blindly imitate the Ford system (one which could minimise the average unit cost by production in large quantities.)' (Monden, 1983, p. 12). Similarly, the changes in process technology which are presently taking place in North America are not just an attempt to replicate the Toyota production system, but rather involve the combining of certain elements of the Japanese production management system ('soft' process technology) with the development of new microelectronic-based 'hard' process technology centred around flexible automation to produce a new hybrid model of production.

Thus both the technical and social organization of production which characterized Fordism are being fundamentally transformed. It is important to distinguish this transformation from the changes in production methods which occurred in the mid-1970s as a managerial response to increasing worker resistance, and which were labelled 'neo-Fordist' by a number of French political economists (Palloix, 1976; Coriot, 1980). Whilst identifying neo-Fordism with some of the elements which also constitute the Toyota production system, e.g. 'autonomous work groups' and electronic process control, these writers seem to view the latter simply as further refinements of Fordist techniques aimed at the further deskilling of workers and the reassertion of *managerial control over* the labour process. Thus, Coriot writes

> the new assembly line is the beginning of a 'new' economy of time and control, which allows for the reproduction of the essential mechanisms of production in the mass production of standardized commodities (a structure necessary for modern mass production) while adapting these to present day conditions in the labour market and to working class resistance. (Coriot, 1980, p. 34)

The discussions of neo-Fordism paid virtually no attention to the organization and management of broader production systems which are such a central feature of the current changes in process technology.

The focus in the remainder of the chapter is on this 'new model of production' which seems to be emerging in the North American automobile industry, emphasizing the technical change that the new model represents when compared to the Fordist model.[4]

Product markets and interfirm competitive strategies

> Henry Ford's soul-destroying, wealth-creating assembly lines are out of date. Most of the things factories make now – be they cars, cameras or candlesticks – come in small batches designed to gratify fleeting market whims. ('The Factory of the Future', *The Economist*, April 5 1986, p. 83)

The late 1970s and early 1980s saw a substantial shift in the market conditions which had led under Fordism to the tendency toward a standardized and stable product design. This shift led the major automakers to modify substantially the 'world car' concept which had been predicated upon the continued and further development of that tendency. Instead they shifted towards a more diverse set of product designs targeted for more regionally conceived markets and a greater awareness of the need for continuous product innovation. It became clear that whilst the boundaries between national markets were breaking down and the world market was converging on certain broadly defined segments, the key competitive strategies within these segments were becoming product differentiation, product innovation, and specialized production targeted at key market niches within the overall market.[5] In response to Japanese competition, product quality along with product innovation became important competitive factors alongside price in the battle for the North American market. These changed market conditions posed a major challenge to the old Fordist production methods,

> even when pursued with greater vigour than in the past, standardization and the pursuit of economies of scale were not enough. Product differentiation and innovation were now essential for maximum sales ...while US methods had focused on economies of scale, Japanese producers had been pursuing a form of production based on the ability to switch readily between models and produce in batches while at the same time minimizing inventories and downtime for machines. (Greater London Council, 1985, p. 249).

Thus in the context of these changed market conditions automakers sought to develop flexible manufacturing systems to maximize their ability to produce efficiently combinations of differentiated products in moderately sized volumes to serve segments of regional and national markets.

Technical organization of the production process

Perhaps the most striking feature of the changes that have occurred in 'hard' process technology has been the large scale introduction of micro-processor-controlled robotics and other CNC machine tools which have had a number of consequences for the car manufacturing process (British Robot Association, 1982). First, the use of these machines results in better product quality through stricter adherence to specifications and manufacturing tolerances, particularly in the early stages of body building. Second, the introduction of these machines has dramatically increased the flexibility of the manufacturing system by reducing the MES for individual product lines and allowing different models to be produced interchangeably on the same assembly line without changing tools or stopping production for long periods of time (Manske, 1982; Bianchi and Calderale, 1983). This flexibility permits rapid adjustments in product mix as market demand shifts and, in particular, will speed up the introduction of new models. Although the capital costs associated with industrial robots are initially very large, these new machines in addition to being labour saving will, in the longer run, also lead to savings of constant capital because they are flexible 'general' purpose machines in the sense that they are reprogrammable.

As was noted earlier perhaps the most important and far reaching consequence of the increased flexibility of 'hard' technology has been the weakening of the link between lowering unit product costs and long production runs based on the achievement of economies of scale. As Altshuler notes 'scale requirements in general are no longer the driving force for industry concentration they have been in the past...modest volumes can be offered at a competitive cost'. (Altshuler, 1984, p. 182). This means that the tendency towards an oligopoly of producers in the world car industry, which was widely predicted in the late 1970s as part of the world car scenario, is now much less likely to occur, and that the future prospects of medium sized and more specialized producers who can establish a dominance in one or more market niches are much enhanced.

Social organization of production

More flexible production systems require more flexible work practices and greater worker commitment to quality control and maintenance. Thus, there has been considerable pressure to change traditional ('Fordist') patterns of shopfloor labour practices and labour-management relations. Hence, a re-composition of skill is taking place at the level of the shopfloor, an attempt to move to broader and more flexible job classifications, and the introduction of techniques such as 'quality circles'. The latter, which are designed to stress individual worker responsibility for quality control and to motivate and involve workers in the constant search for improvements in production methods, also reduce the need for the army of foremen, supervisors and checkers on the shop floor which characterized Fordist production. 'Quality circles' and the use of statistical quality control techniques among suppliers have been introduced in an effort to reduce defective parts to an absolute minimum. The latter increases process yield and reduces the size of the required inventory buffers which were formerly required to achieve high process yield. The old adage that quality entails increased production costs has been stood on its head.

In addition to the changes in production organization within the assembly plants, the early 1980s also witnessed a restructuring of the relationships between the assemblers and their component suppliers. These changes include an increase in the outsourcing of parts by the auto assemblers; the elimination of multisourcing of parts in favour of single sourcing, a reduction in the number of suppliers, and longer term contracts with suppliers, all designed to improve productivity among suppliers; and, the development of closer working relations with the component suppliers in order to involve the latter in ongoing product design and development and to encourage them to improve quality control (Holmes, 1986a, pp. 96–8).

However, perhaps the most significant change, certainly in terms of its potential future impact on the geography of the industry, is the development of just-in-time (JIT) delivery systems (often and somewhat erroneously conflated with the kanban method of stock control) which were pioneered by Toyota in Japan (Sugimori, 1977; Sheard, 1983; Monden, 1983; Estall, 1985). The timing and coordination of the delivery of parts from the component suppliers and subcontractors to the assembly plant play an integral role in the functioning of auto production systems in Japan.

By reducing to a minimum the inventory levels for the production system as a whole, the JIT system can result in significant improvements in the rate of accumulation by reducing the amount of required circulating capital and, thus, reducing the turnover time of capital. For example, Sheard notes that one Japanese study estimated that in one year Toyota saved the equivalent of 10 per cent of its net profit through the operation of its JIT system (Sheard, 1983), p. 62), and Sugimori (1977) provides startling figures on the reduction of capital turnover times achieved by Toyota and other Japanese companies. Note also that whilst the operation of a JIT system depends on strict quality control it, in turn, contributes to improved quality control by pinpointing quality problems in deliveries from suppliers.

Whilst it is unclear as to just how close North American automakers can come to emulating the almost zero inventory levels achieved in Japan because of the differences in geographical scale and the industrial and locational inertia embodied in existing production systems, '. . . the trend [in North America] towards the recentralization of the production of major components around each production location is inevitable if the "just in time" system is to be fully developed on the Japanese model.' (Jones, 1985, p. 160). The fact that MESs are beginning to fall in component production will facilitate the replication of component production facilities at each major assembly site.

Therefore, the impact of these developments on the locational structure of the auto industry in both Canada and the US will likely be to eliminate suppliers in more peripheral areas and to concentrate final production and assembly in areas with well developed networks of suppliers (i.e. a reconcentration of production into the traditional auto making regions of the Upper Midwest and Southern Ontario). Thus major new investments in South Central Ontario were announced in late 1985 and early 1986 by Honda, Nissan, AMC–Renault and GM. GM is investing $2 billion in the design and building of GM Autoplex at Oshawa '...a fully integrated, totally synchronized manufacturing system, tying together our resources and those of our suppliers into one of the most modern automotive manufacturing complexes in the world' (*Toronto Star*, March 25 1986, p. E1). One of the key elements of the project, which will create at Oshawa the most advanced and largest vehicle assembly complex in North America, is an explicit move to just-in-time delivery. Forty per cent of parts will be made within four hours travelling time of Oshawa and for the first time major body panels will be stamped at Oshawa rather than imported from the US. It is significant that the reports of this announcement also stressed the uncertainty that exists regarding the future of the GM assembly plant at Ste Therese, Quebec, a plant whose location becomes increasingly more peripheral relative to the main locus of the North American auto industry (*Toronto Star*, March 25 1986, p. E1; *Globe and Mail*, March 25 1986, p. B1).

The standard Fordist labour contract which set wage increases and regulated labour relations in the North American auto industry from the late 1940s until the end of the 1970s is also in the process of being radically transformed (Katz, 1984). Labour contracts negotiated by the UAW in 1985 and, in particular, those for the Mazda assembly plant at Flat Rock, Michigan, the GM Saturn project at Spring Hill, Tennessee and the GM–Suzuki joint venture in Ontario, established a new model for collective agreements in the North American auto industry. This new model includes a significant reduction in the number of job categories and much broader job descriptions designed to provide flexibility in the assignment of workers to tasks, and the abolition of the old Fordist formula wage rules, instead linking wage increases to performance and wage rates to seniority. For example, in the instance of the GM–Suzuki project '...Suzuki wants concessions on wages and worker flexibility never before granted by the UAW in Canada...one possibility would be to use an existing plant but both companies [Suzuki and GM] seem to prefer an all-new plant with a work force suited to Suzuki's needs...in separate talks with UAW–Canada Suzuki is known to be seeking slightly lower wage levels than the current industry average, fewer classifications and greater flexibility.' (*Globe and Mail*, February 7 1986, p. B3).

These changes in the social organization of production add up to a radical change in shopfloor labour practices. However, they require not only a change in attitudes and practices on the part of workers, but also changes in management practices. The initial management response to the crisis was to attempt to increase the rate of exploitation by reducing the number of paid holidays, revising work practices and rolling back wages by breaking the pattern of annual increases in base wage rates which, since the late 1940s, had been a standard feature of the employment relation under Fordism. But as Littler and Salaman, writing about similar changes in Western Europe, note:

> the 1980s like the 1930s will no doubt spur some employers to cut wages, extend work hours in some form or impose straightforward labour intensification involving no attempt at work reorganization...but there is a limit to what can be achieved through economic coercion. In particular it ignores the changing nature of competition. An emphasis on quality rather than the velocity of throughput means that reluctant acquiescence has to give way to active cooperation...the crisis of Fordism is neatly exemplified by the present dilemma of Ford Europe. Faced with the large productivity gap between its European plants and Japanese manufacturers, Ford set up its 'After Japan' programme which as the headlines announced was a contradictory mix of robots, job cuts, a union struggle *and* an attempt to engage the enthusiasm and willingness of the workers! (Littler and Salaman, 1984, p. 90).

It has already been noted that because of recent technical change, the auto industry is likely to remain geographically concentrated in the traditional auto producing regions of the Upper Midwest and Southern Ontario, and it has been emphasized that the transformation of production which is underway is not simply a technical process but is one that also involves the social reorganization of production within the plant. This means that contrary to Clark's argument that '...the previous pattern of labor agreements have paralyzed firms' capacities to adjust production strategies in the midwest region. As a result, regional switching has taken the place of in situ adjustment.' (Clark, 1986, p. 137), the automakers are now faced with the task of remaking shop floor labour practices in situ. Whilst the 'running away from the unions' strategy described by Clark was characteristic of the 1970s, when there was a shift in new investments in the US auto industry away from the midwest to 'greenfield' sites and 'green' labour markets in the sunbelt, his analysis fails to recognize that this strategy has been largely eclipsed by the technical change which has occurred in the industry in the 1980s and, in particular, by the tremendous pressure exerted by the adoption of JIT methods for the reconcentration of auto production in the midwest.

However, the work practices cannot simply be remade and implemented by managerial fiat, they are likely to be the object of struggle on the shopfloor, the outcome of which is likely to be very uneven. Management seems to have encountered far greater resistance to their attempt to implement some of these strategies in Canadian auto plants than in US plants. On at least three occasions – union resistance in Canada to worker concessions as part of the Chrysler rescue, the 1982 strike against Chrysler Canada to restore wage increases, and the 1984 strike against GM Canada which was successful in preserving the traditional post-war Fordist model of annual base wage rate increases whilst at the same time pressing (less successfully) for a reduction in working time – Canadian autoworkers have demonstrated their willingness to adopt a more resistant and militant stance than their counterparts in the US. Not surprisngly this has created tensions over bargaining strategy within the UAW and, in 1985, ultimately led to the secession of the Canadian autoworkers from the UAW to form their own autonomous organization, the UAW–Canada.

In large measure, the different bargaining posture assumed by the UAW in the two countries can be explained by the differential employment impacts produced by uneven development within the North American automotive industry (figure 7.3). The extent of job losses and plant closings in the US auto industry has dramatically undermined and weakened the bargaining strength of the UAW–US and has made issues such as job security and extended unemployment and retraining benefits of prime importance in recent contract negotiations. By contrast, the employment impact of the crisis on the auto industry in Canada has been much milder, and in recent years Canada has once again enjoyed a considerable cost competitive advantage over the US. Obviously one key factor which affects levels of automotive production in Canada is the currency exchange rate with the US. Although there has been nominal wage parity between Canadian and US auto workers since the early 1970s, the exchange rate since 1978 has given Canada a considerable labour cost advantage. For example, at the time of the labour contract negotiations with GM in 1984 it was acknowledged that there was a \$C7.50 an hour (the base hourly wage rate under the previous contract had been just over \$13.00 for an assembly line worker) labour cost differential in Canada's favour due to the exchange rate and differences in the cost of fringe benefit packages. This helps to explain why the auto industry recovered much more strongly in Canada than in the US from the 1979–82 downturn, so much more strongly, in fact, that between 1982–4 Canada recorded an accumulated surplus of over $12 billion on the automotive trade with the US under the Auto Pact with a further $6 billion expected to be added to the surplus in 1985. Thus, from this position of relative strength the UAW–Canada has been able to fight to maintain or enhance the share of value added going to workers, and has sought to protect jobs in the face of the significant gains in productivity which are being made through the application of new process technologies, by pressing for a reduced work week with no loss of real income.

CONCLUSION

This chapter has focused on the very rapid changes in production technology which are occurring in the North American auto industry and, whilst it is still much too early to assess their full impact, has tried to sketch out some of the consequences of these changes for the organizational structure and geography of the auto industry in Canada. It has also tried to demonstrate that a new model of production is emerging in the auto industry which represents a sharp and distinct break from the model of production which characterized Fordism, the regime of accumulation which was manifested in the post-1945 boom. In fact, it was the very crisis of Fordism in general, and of Fordist techniques of production in particular, which triggered the period of restructuring from which this new model of production is emerging. In contrasting the production of automobiles under the new model with their production under Fordism, the strong and mutually reinforcing inter-relationships that exist within each model between the nature of product markets has been emphasized together with the competitive context within which inter-capitalist competition occurs, the organization of the production process, in both a technical and social sense, and both within individual plants and between the plants which constitute a production system.

In the early phases of the crisis most studies of the international auto industry were predicated on the assumption that the solution to the crisis lay simply in extending the Fordist model of production to the global level. These studies were based on the conviction that an increasing degree of product standardization was inevitable, and that the key competitive strategy for automakers would remain one of lowering unit production costs through increased automation and economies of scale. Thus, the need to cut costs indicated the further internationalization of production, a shift of production to low wage locations, and the development of the 'world car'. The uncritical acceptance of this scenario led several writers to predict an extremely bleak future for the Canadian auto industry (e.g. Perry, 1982; Van Ameringen,

1985). However, the 'world car' strategy, which was so widely touted in the late 1970s as the solution to the crisis in the auto industry, is now rejected by most industry analysts as the most liekly future scenario for the world auto industry. Instead they argue that

> the new production technologies mean that the shift to low wage locations will not occur on the scale once expected. The markets of the developed countries are demanding precise, high quality production. Flexible manufacturing in combination with the redesign of products to gain its full benefits can provide this while sharply increasing labour productivity. These innovations have shifted the focus of thought about the future geography of production location from the less developed countries to the concentrated production of most components near the point of final assembly in the developed countries. (Altshuler, 1984, p. 249).

If this tendency proves to be more enduring than that towards the 'world car' (which seems likely) then the short and medium term prospects for the survival of a significant auto industry in Canada are much enhanced.

One thing that this chapter has made clear is that the solution to the crisis can not and will not be achieved through a return to the 'golden age' of Fordism. The restructuring crisis which now has gripped the world economy for over a decade involves (but as has been stressed throughout the chapter is not confined to) substantial changes in production technology, both with respect to the shop floor organization of the labour process and the broader organization of production systems. The latter, in particular, is likely to have a profound impact on the locational structure of industry (and it might be argued on location theory too!).

Whilst the auto industry appears to be once again at the forefront of the introduction of radically new process technology, the latter is also having a profound economic and social impact on firms and workers in many other industries. All the evidence suggests that these circumstances represent a critically important moment, not only in the history of the development of the forces of production but also in the history of human development, since

> at present we seem to be in a phase of development where the 'new' technology is genuinely new, that is, it appears to be malleable and to offer a range of options – centralization versus decentralization, enhancement of skills versus polarization of skills away from the shop floor, rigid job control versus delegation of decision making over production. (Littler and Salaman, 1984, p. 97).

It is important to struggle to ensure that the 'correct' choices are made and that the new technology is used in a progressive and liberating manner rather than one which is repressive – although in the absence of a thorough transformation of the social relations of production the odds remain heavily weighted towards those options favoured by capital in its relentless quest for surplus value.

NOTES

Grateful acknowledgement is made to the Social Sciences and Humanities Research Council of Canada who provided funding which enabled me in August 1985 to attend the 3rd Anglo-Canadian Symposium on Industrial Geography held at Swansea, Wales and the IGU Commission on Industrial Change Symposium held at Nijmegan, The Netherlands at which earlier versions of this chapter were presented as a paper. Figures were prepared for publication in the Department of Geography, Queen's University by Ross Hough and George Innis.

1 In this chapter the terms 'automobile industry' and 'automotive industry or sector' are used interchangeably and include both the final assembly of passenger vehicles. Whilst the analysis focuses on the 'Canadian' automotive industry it is, of course, impossible to discuss the latter in isolation since automotive production within Canada and the United States have been organized as one integrated production system since the mid-1960s. However, as we shall see it is misleading to assume that developments in the 'Canadian' portion of the North American auto industry simply mirror or echo those in the 'US' industry.
2 The process yield of a given labour process is determined not only by line speed but also by the number of defective parts produced and by the percentage of nonproductive 'downtime'.
3 Quality circles and the kanban system for conveying information about stock levels are just two elements of the overall Toyota production system. The most comprehensive account of the latter available in English is Monden (1983).
4 The term 'new model of production' is used with extreme caution. Only time will tell whether the developments in production technology described here will truly form the basis for a new and stable period of sustained accumulation.
5 The fragmentation and segmentation of the mass market in the 1980s is not confined to the market for automobiles; it is clearly evident in the clothing, food retailing and catering industries. Sabel (1983) argues that it is the major factor contributing to the crisis of Fordism since the destruction and fragmentation of the mass market removes a necessary conditions for the development and use of Fordist mass production technology.

REFERENCES

Abernathy, W. (1978) *The Productivity Dilemma: Roadblock to Innovation In the Automobile Industry*, Johns Hopkins Press, Baltimore.

Aglietta, M. (1979) *A Theory of Capitalist Regulation: The US Experience*, New Left Books, London.

Altshuler, A. et. al., (1984) *The Future of the Automobile: The Report of MIT's International Automobile Program*, MIT Press, Cambridge, Mass.

Armstrong, P., Glyn A. and Harrison J. (1984) *Capitalism Since World War II*, Fontana, London

Bianchi, S. and Calderale, I. (1983) Flexible Manufacturing and Product Differentiation in the Automobile Industry, Working Paper I–B–83–2, MIT Future of the Automobile Program.

British Robot Association (1982) *Proceedings of Robots in the Automotive Industry: An International Conference*, IFS Publications, Bedford.

Canada, Department of Regional Industrial Expansion (1984) *1981 Report of the Canadian Auto Industry*, DRIE, Ottawa.

Canada, Federal Task Force (1983) *An Automotive Strategy for Canada*, Minister of Supply and Services, Ottawa.

Clark, G. L. (1986) The crisis of the Midwest auto industry. In Scott, A. J. and Storper M. (Eds.) *Production, Work, Territory: The Geographical Anatomy of Industrial Capitalism*, Allen And Unwin, London.

Clark, P. and Tann J. (1986) The Translantic Diffusion of the Assembly Line: The Case of the Automobile Industry. Paper presented to the Aston/UMIST Labour Process Conference, Aston University, Birmingham, April 1986.

Cohen, R. B. (1983) The new spatial organization of the European and American automotive industries. In Moulaert F. and Salinas P. W. (Eds.) *Regional Analysis and the New International Division of Labor*, Kluwer-Nijhoff Publishing, Boston, 135–44.

Cole, R. E. and Yakushiji, T. (1984) *The American and Japaness Auto Industries in Transition*, Center for Japanese Studies, University of Michigan, Ann Arbor.

Coriot, B. (1980) The Restructuring of the assembly line: a new economy of time and control, *Capital and Class* 11, 34–43.

Dankbaar, B. (1984) Maturity and relocation in the car industry, *Development and Change* 15, 223–50.

Davis, M. (1984) The political economy of late Imperial America, *New Left Review* 143, 6–38.

Davis, M. (1985) Reagonomics' magical mystery tour, *New Left Review* 149, 45–65.

De Vroey, M. (1984) A regulation approach interpretation of the contemporary crisis, *Capital and Class* 23, 45–66.

Estall, R. C. (1985) Stock control in manufacturing: the Just-in-Time System and its locational implications, *Area* 17, 129–33.

Evans, D. and Kaplinsky, R. (1985) Slowdown or crisis? Editorial introduction, *IDS Bulletin*, 16, 1–8.

Fraser, J. (1982) Changes in production process in the automotive industry, Working Paper, US–C–82–1, MIT Future of the Automobile Program, Cambridge, Mass.

Friedman, D. (1983) Beyond the Age of Ford: the strategic basis of the Japanese success in automobiles. In Zysman J. and Tyson, L. (Eds.) *American Industry in International Competition*, Cornell University Press, Ithaca.

Frobel, F., Heinrichs, J and Kreye, D. (1979) *The New International Division of Labour*. Cambridge University Press, Cambridge.

Fulco, L. J. (1984) Strong post-recession growth in productivity contributes to slow growth in labor costs, *Monthly Labor Review* Dec., 3–10.

Gartman, D. (1982) Basic and surplus control in Capitalist machinery: the case of mechanization in the auto industry, *Research in Political Economy* 5, 23–57.

Greater London Council (1985) *The London Industrial Strategy*. Greater London Council, London.

Hawkesworth, R. I. (1981) The rise of Spain's automobile industry, *National Westminster Bank Quarterly Review* February, 37–48.

Holmes, J. (1983) Industrial reorganization, capital restructuring and locational change: an analysis of the Canadian automobile industry in the 1960s, *Economic Geography* 59, 251–71.

Holmes, J. (1986a) The Organization and locational structure of production subcontracting. In Scott, A. J. and Storper, M. (Eds.) *Production, Work, Territory: The Geographical Anatomy of Industrial Capitalism*, Allen and Unwin, London.

Holmes, J. (1986b) The crisis of Fordism and the restructuring of the Canadian auto industry. In Holmes J. and Leys C. (Eds.) *Front Yard/Back Yard: The Americas in the Global Crisis*. Between the Lines Press, Toronto (forthcoming).

Holmes, J. and Leys. C. (1986) Introduction. In Holmes J. and Leys, C. (Eds.) *Front Yard/Back Yard: The Americas in the Global Crisis*. Between the Lines Press, Toronto (forthcoming).

Jenkins, R. (1984) Divisions over the international division of labour, *Capital and Class* 22, 28–58.

Johnston, W. (1982) Issues in the Multinational Sourcing of Production, Working Paper, US–B–82–6, MIT Future of the Automobile Program, Cambridge, Mass.

Jones, D. T. (1985) Vehicles. In Freeman C. (Ed.), *Technological Trends and Employment: 4. Engineering and Vehicles*, Gower, Aldershot, 128–187.

Katz, H. C. (1984) The US automobile collective bargaining system in transition, *British Journal of Industrial Relations* 22, 205–17.

Lewchuk, W. (1983) Fordism and British motor car employers. In Gospel, H. and Littler, C. R. (Eds.), *Managerial Strategies and Industrial Relations*, Heinemann, London.

Lipietz, A. (1982) Towards global Fordism? *New Left Review* 132, 133–47.

Lipietz, A. (1985) *The Enchanted World: Inflation, Credit and the World Crisis*, Verso, London.

Lipietz, A. (1986), The World Crisis: The globalization of the general crisis of

Fordism. In Holmes, J. and Leys, C. (Eds.) *Front Yard/Back Yard: The Americas in the Global Crisis*, Between the Lines Press, Toronto (forthcoming).

Littler, C. R. (1982) *The Development of the Labour Process in Capitalist Societies*, Heinemann, London.

Littler, C. R. and Salaman, G. (1984) *Class at Work: The Design, Allocation and Control of Jobs*, Batsford Academic, London.

Mandel, E. (1980) *Long Waves of Capitalist Development*, Cambridge University Press, Cambridge.

Manske, F. (1982) Economic efficiency and flexibility of IR application in automobile production. In British Robot Association (1982), 31–38.

Monden, Y. (1983) *Toyota Production System*. Industrial Engineering and management Press, Norcross, Georgia.

Morgan, K. and Sayer, A. (1984) A Modern Industry in a Mature Region: Electrical Engineering in South Wales. Urban and Regional Studies Working Paper, No. 39, University of Sussex.

National Academy of Sciences (1982) *The Competitive Status of the US Auto Industry*. National Academy Press, Washington.

Norsworthy, W. and Malmquist, D. H. (1983), Input Measurement and Productivity Growth in Japanese and US Manufacturing, *American Economic Review*, 73, 947–67.

O'Connor, J. (1982), The meaning of crisis, *International Journal of Urban and Regional Research* 5, 301–29.

Palloix, C. (1978), The labour process from Fordism to Neo-Fordism. In *The Labour Process and Class Strategies*. Stage 1, London, 46–65.

Perry, R. (1982) *The Future of Canada's Auto Industry*, James Lorimer, Toronto.

Sabel, C. (1982) *Work and Politics: The Division of Labour in Industry*, Cambridge University Press, Cambridge.

Sheard, P. (1983) Auto-production systems in Japan: Organizational and locational features, *Australian Geographical Studies* 21, 49–68.

Stone, K. (1981) The Post-war paradigm in American labor law, *Yale Law Journal* 90, 1509–80.

Sugimori, Y. et. al. (1977) Toyota production system and kanban system: materialization of Just-in-Time and Respect-for-Human System, *International Journal of Production Research*, 15, 553–64.

Transnational Information Exchange – Europe (1983), *Left-Hand Drive: Shopfloor Internationalism and the Auto Industry*, TIE Europe, Amsterdam.

UNCTNC (1983) *Transnational Corporations in the International Auto Industry*. New York: United Nations.

US Department of Transportation (1981) *Auto Industry 1980*, DoT Report P–10–81102, Cambridge, Mass.

Van Ameringen, M. (1985) The restructuring of the Canadian automobile industry. In Cameron, D. and Houle, F. (Eds.), *Canada and the New International Division of Labour*, University of Ottawa Press, Ottawa, 267–287.

Ward, M. F. (1982) Political economy, industrial location and the European motor car industry in the postwar period, *Regional Studies* 16, 443–53.

Zohar, U. (1982) *Canadian Manufacturing: A Study in Productivity and Technological Change*, James Lorimer, Toronto.

8
Technical change and the restructuring of the North American steel industry

J. H. Bradbury

In order for most mature industries to remain competitive they must undergo a constant pattern of adaptation to changes in the conditions of production. These modifications may be within the actual machinery used to produce commodities, in the plant itself, involving changes to the relationships between different systems of production, or in the methods whereby labour is deployed in the production process. All these can be considered as internal responses to changes in the production system *per se*. As such they are under the direct control of both workers and management and can be manipulated to alter the rate, type and function of production (Harvey, 1982).

To these internal responses must be added those external conditions which impinge upon the viability and competitive status of a production system. Most, if not all, industries exist within a regional, national and inter-national system. World fluctuations in prices, commodity overproduction and underconsumption, stockpiling and the internationalization of capital and production, have created conditions which impinge upon the competitive relationships of a local plant or industry. Furthermore they are subject to changes in conditions and phases in the business cycle: in periods of expansion and upswing, competition is marked by growth; in downswings competition is moderated by retrenchment and restructuring (Bradbury, 1984; 1985). In phases of industrial restructuring these external constraints condition the capacity of the local industry to survive, grow and remain competitive.

It is clear that an industrial production system consists of a close and almost organic relationship between the level and rate of production, the use of technology and the deployment of labour. It is equally clear that decisions made in one area of production are influenced by, and will have a direct bearing, upon, the allocation of resources in other areas of the production system. In its simplest terms, fixed and variable capital are the primary sources of variance in a production system. It is here that are found the major areas of conflict over policy in a phase of restructuring or indeed at all times during the production process. In periods of rapid restructuring, especially during a crisis phase, both fixed and variable capital will be reformed and reallocated. In some instances this will mean the shifting of machines, closing down or opening of new plants, slowing or speeding of machine rates and the reallocation of labour – either to new plants, new machines or to unemployment.

Restructuring in downswing periods in the mature industries generally has a negative impact upon workers and their families. The loss of jobs, decline in community viability, migration and the insecurity brought about by industrial winding down is too often overlooked in the literature. Indeed, the human consequences of capital restructuring are a direct outcome of the changes in the balance of fixed and variable capital in an industrial system. In such instances, technological change in a mature industry can have a number of effects. First, the new technology may introduce plant-saving techniques to make a plant efficient, at least in the short term, and thus retaining the existing level of labour. Second, the move towards incremental substitution of labour by capital (fixed and variable capital), may result in a slowing down of new employment and a rise in productivity. Third, the wholesale substitution of labour by new technology, may result in job losses, technological substitution of previously labour-intensive operations and the entry of the plant and production system into a completely new capital to labour ratio. In practical terms, all forms of technological substitution can occur at different stages of the life cycle of a plant depending upon the pressures of class conflict, industrial competition and the access to invest-ment capital for research and development (Massey, 1982).

Technology, *per se*, is part of the direct allocation of machinery and fixed capital in the production process. It is, therefore, a major variant in the agenda of restructuring both during normal phases of production and during periods of upswing and downswing. There is sometimes an implicit assumption that 'mature' industries are technologically stagnant. Much of the recent analytical work in this field has failed to recognize that change in such industries can have more far-reaching economic, social and geographical consequences than advances in high-technology sectors.

A large number of mature or sunset industries have undergone some dramatic changes especially as a result of the major international recession of the early 1980s. During this recession, plants were closed, jobs were lost (or gained) and major areas of traditional industry were closed down and abandoned. Not all these changes occurred because of a lack of a modernization strategy and the required investments. Some changes occurred because of long term disinvestment and reallocation of resources by large manufacturing and processing organizations. Other changes occurred in parallel with the internationalization of capital and the movement of corporations to tax havens, to low wage areas or to new agglomerations of technological expertise.

Technological change *per se* is, therefore, only one of a number of variants in both the cause and response to industrial restructuring. Technology in this instance is referred to as the content and input of machinery, operating systems, research and design and technical aids to existing and future production systems. It embraces the area of scientific and engineering skills and the application of knowledge to producing, maintaining and improving a system of production. In most instances this knowledge becomes embodied in the production machines of a plant and is then deployed to create the basis for new machines which may either improve upon or even replace those machines or plant. It is, therefore, imperative to mature industries, as well as to others, to maintain a significant input into this phase of production if the plant is to run, be repaired and maintained and operate competitively.

In the simplest terms, a plant may remain competitive, provided that labour costs and other inputs and supplies of variable capital are controlled by modifying the speed at which production occurs: that is to alter the rate at which the existing fixed capital is used. New technological innovations may, however, be introduced into an existing plant to improve the efficiency and running rate of either a portion of a machine, a whole machine or a whole production system. Technological innovation, therefore, is constantly sought by firms to improve the efficiency of the existing fixed capital. The addition of often simple devices to existing machines may enable production to be speeded up, labour hours modified or workers replaced or reallocated. Such innovations may occur continually in a plant and they may achieve both minor and major changes in the rates and quality of production. Indeed, in many older or 'mature' industries this is often the medium whereby new technology is introduced.

The rate at which innovations occur and the methods of deploying or even obtaining exclusive rights to them, are vital to the survival of some mature industries during phases of critical restructuring. Competition for innovations in technology at such times is generally high and the cost of keeping up with their potential impact on the production system itself is also high. Those firms which cannot afford to restructure because of the high cost of access to innovations may not survive the crisis. Occasionally they may survive by some form of in-house modification to their production system (local inventions and innovations), but for the most part their access to industry-wide technological inputs and innovations is limited either by their lack of access to capital for investment or by the caution or technical ignorance of local management and engineering staff.

In all fairness to local managers, however, the channels to both small-scale and large-scale technology in a phase of restructuring may be severely limited by the events which trigger a crisis and the mechanisms and investments available to modify the balance of either fixed or variable capital. In many larger and older (mature) industries, the actual costs of investing in new plants may be very high so that the very volume of capital required may be prohibitive in a restructuring phase. Also the timing of a downswing may be the occasion on which firms are least likely to have access to liquid assets for investment in new technology, in replacement of parts and machines, or even in repairs and maintenance. Furthermore, the frequency of crises may mean a lowering of a firm's capacity to survive a restructuring crisis.

Decisions to invest in new technology, new plant and new systems of production are made on the basis of expected and planned survival or profitability in the future. In many 'mature' industries requiring large amounts of fixed capital investment, such decisions can only be made on the basis of long term projections. A decision to build a steel mill for instance, may be made up to ten years or so before its actual construction. Therefore, both 'greenfield' and 'brownfield' site constructions may be exposed to the vagaries of price fluctuations and upswing and downswing in the international production system. Those mature industries with an expected life span and amortization period of more than, say, twenty years, can thus expect to be subjected to several crisis periods which will devalue or undermine their initial investments. Indeed, restructuring crises may envelope an apparently viable system of technology and bring about its demise.

For the most part technological changes may parallel restructuring crises, or they may occur at random throughout the events of either a business cycle or the expected life span of a plant. A technological 'miracle' cannot be expected to emerge when one is needed, although sometimes in the 'mature' industries small 'modernizations' have been known to permit a plant to survive a crisis period of restructuring.

TECHNICAL CHANGE AND RESTRUCTURING IN THE
INTERNATIONAL STEEL INDUSTRY

The steel recession between 1981 and 1983 was a traumatic one in North America, as it was elsewhere (Barnett and Schorsch, 1983; Hogan, 1983; Hudson and Sadler, 1983). Major corporations collapsed, were restructured, rationalized, wrote-off plant and experienced heavy financial losses. Several thousand workers were temporarily laid off or lost their jobs permanently. At the same time the trauma produced a 'leaner look' in several major firms focusing on different markets. It was accompanied by a switch of investment to new plants and to new technology in smaller mini- and midi-mills and to non-steel industries. The intensity of the process was beyond a normal cyclical event (Allan, 1984); it embraced most national steel industries, albeit unevenly. The global rupture of steel producers and the decimation of parts of its labour force undermined the working class culture and vitality of a number of communities and brought the international steel industry to a sudden reassessment of its fragile condition. In Canada and the United States there was a two to three year decline in production, a general loss of revenue and a 23 per cent and 40 per cent decline, respectively, in steel employment. At the same time, the uneven impact of the recession left a number of regional steel mills with little margin to survive, closed others, but allowed some to engage in new investments in technology in both 'greenfield' and 'brownfield' sites.

In retrospect, the recession of the early 1980s constituted something more than a cyclical downswing and upswing designed to correct supply and demand in a recognizable economic cycle. The recession was clearly not limited to single sectors, such as steel, but was interlocked with a series of trigger events in different sectors dating back to the early 1960s. For a period of some twenty years there was a potential for crisis, but it was only the conjuncture of a number of trigger points which resulted in the recession of the 1980s. Clearly the events preceding the downswing reflected major structural alterations in the global economic system. The increase in the rate of the internationalization of capital and the advent of 'sunrise' industries, often to the disadvantage of 'sunset' industries, was an integral part of the restructuring of the global industrial system (Bradbury, 1985). There was a shift in the leading industrialized nations away from heavy manufacturing toward different forms of assembly, high-technology, information and service industries, as well as an accompanying formula of rapid industrialization in developing countries.

The changes in the global system had a variable effect on the steel industry. The impacts have been most severe in Western Europe and North America, where long-term declines have resulted in closures and ruptures in regional industrial systems. Why did this take place? Was it an inevitable outcome of structural and systemic problems? Was it predictable? It is important to recognize that some reasons for the demise of heavy and capital-intensive industry lie in events in the early 1960s. Up to that time the United States steel industry was dominant in both a technical and a market sense.

War-battered steel industries in Europe and Japan began the long process of rebuilding and modernizing. In so doing they laid down the basis of a more competitive form of behaviour and opened channels for new technology in steel making. In the late 1960s these firms were joined by a new group of steelmakers in emerging Third World countries. The acquisition of a national flag, an airline and a steel mill was on the development agenda of a number of these nation states (Allan, 1984). The conjuncture of the result of this massive increase in productive capacity with the most severe slump since the 1930s and the restructuring of the new international division of labour, contributed toward the traumatic environment in the steel industry of huge operating losses, retrenchment, plant closures and massive temporary and permanent layoffs. To this must also be added a number of trigger events which influenced the timing and the uneven impact of the recession. Undoubtedly the energy-cost spiral, initiated by changes in oil prices, boosted inflation and increased the input costs to steel mills. Raw-material costs also rose and access to increasingly competitive markets had a significant backlash on iron mining regions (Bradbury and St. Martin, 1983; Bradbury, 1984). The intensification of competition, especially in the production and marketing of semi-finished steel from developing countries, meant that pressure was placed on traditional producers and on ageing machinery in Europe and North America.

Coincident with these problems and exacerbating the restructuring trend in the recession of the early 1980s, there was a shrinkage in steel markets for some commodities. This can be seen in the collapse of steel export markets and a decline in steel intensity measures. The ratio of steel consumption to GNP in the industrialized world has been declining for some years. For example, steel intensity in the USA has declined from 90 tons/million dollars of GNP in 1970 to 54.4 tons in 1984. This occurred as a result of a number of changes in industries consuming steel: the quantity of steel used by the auto industry for instance declined by more than 40 per cent between 1979 and 1982. Changes in the construction industry after 1965 resulted in a 45 per cent decline in steel useage (from 34 lb/ft^2 in 1965 to 19 lb/ft^2 in 1984) (Allan, 1984). Declining steel intensity also resulted from the intervention of competing materials including aluminium, plastics and light-weight alloys.

The decline of markets was exacerbated in some instances by overproduction and in others by short-term destocking by steel consumers. As the recession deepened, destocking continued back to the point where inventories were so low they affected the capacity of steel mills to operate and resulted in closures of linked supply industries.

RESTRUCTURING CONCEPT APPLIED TO THE STEEL INDUSTRY

Within the steel industry there are a number of specific structural and geographical components to consider. In the short run, restructuring, especially in a crisis, embraces significant job losses and disruption to workforces and communities dependent upon the steel industry. Whereas low demand may be accompanied by operational rationalization to minimize costs, which results in the closure of facilities, many permanently, a general restructuring crisis can be overcome by increased productivity and improvements in international competitiveness through better facilities and techniques and the development of new products (ISI, 1983a and 1983b). There are both positive and negative regulators of steel costs which affect output and which can be altered in a restructuring phase. These may be understood as investment mixes and decisions which alter the balance of funding in relation to machinery, raw materials and labour (Harvey, 1982).

The timing of new investments is related to the phase of technological change and to critical points in the success (or failure) of the production process. Major retooling can take place at the bottom of a downswing or a recession, because it is cheaper to do so (*Iron Age*, 1984b). In a general sense, periods of crisis are also periods of dramatic restructuring. The forms of restructuring evident in expansionary phases, however, tend to fill in the industrial patterns and sequences set at an earlier period – often during the crises which sparked the restructuring phase (Smith, 1984). Some major steel firms show strong tendencies and even a preference for technological change in the bottom of a down cycle. This is because labour-saving devices may be introduced more readily at the bottom of a cycle, whilst new technology directly related to more efficient production may come in a upswing (Aylen, 1983, 1983b; Kawahito, 1982). Major technological expansions, as in Japan in the 1960s, were instituted at the beginning of an upswing, to take advantage of technological breakthroughs, including the basic oxygen furnace, giant ore carriers, computers and continuous casting equipment (*ISI*, 1982).

Revamped technology in the international steel industry influenced the restructuring processes available to firms in the 1970s and 1980s. In addition, the industrial make-up of advanced economies tended to move away from traditional steel-using sectors toward advanced technology including electronics, computers, instruments, armaments and aerospace. Indeed, as noted earlier, there was a relative steel shrinkage in advanced capitalist economies (Warrian, 1984). At the same time, steel use and production grew in New Industrializing Countries (NICs) and oil rich regions. Countries such as South Korea, Taiwan and Brazil, part of the new wave of steelmakers, became the aggressors in the international market. It is from these new producers that competition at the international level generated the conditions under which latter day restructuring took place (*ISI*, 1983a).

It is imperative to recognize that industrial restructuring *per se* is both a contemporary and historical phenomenon. It is not a new or recently discovered process; rather, cyclical and crisis behaviour are normal within different fractions of capital (Harvey, 1982; Moulaert and Salinas, 1983). While the restructuring of the 1970s and 1980s reached dramatic proportions in North America and Europe (Warrian, 1984), there have been phases in the past in which equivalent periods of concentrated change have occurred in the steel industry. Indeed, restructuring, in and of itself, is an on-going process in which industrial capital strives to maintain an equitable balance between fixed and variable capital (Harvey, 1982). Any surge or sag in this balance will lead to uneven patterns of accumulation or to industrial decline. The level of adjustment required in the restructuring is, therefore, a measure of the degree of imbalance in investments or the level of competition in production, costs and technology.

While infrastructural and labour process change are integrated within the plant and production itself, political processes may act upon the specific economic conditions at any one time or place of production. The state may take various actions either to encourage growth or to manage crises. In this category come tariffs, government aid, protectionism and the facilitative role of the state in shaping work conditions and labour relations. Here too it is necessary to broaden the role of the state to include various forms of trade and market control (Clark and Dear, 1984), including regulation of production quotas, prices and capacity. For instance, steel companies in the EEC are subject to a significant degree of regulation. In keeping with government rationalization plans, most major integrated producers in the EEC have reduced or plan to reduce, their raw steel-making capacity (Hogan, 1983). In Canada, as will be shown, government activities embrace investments in production facilities through ownership of equity capital, tariff structures and anti-dumping laws.

RESTRUCTURING DURING THE RECESSION IN THE UNITED STATES

The steel industry in the US began to shrink in 1977 but even before then there were signs of upheaval and conflict over the access of imports and over strikes. In 1959, the first major period of imports occurred in conjunction with a series of strikes (Barnett and Schorsch, 1983). In the 1960s, steel mills expanded only slowly against a rise in imports. However, between 1968 and 1974, government responded to pressures against an open market policy and provided some local protection. Despite this the steel industry could not cope with the foreign competition and began to lag behind. By 1974 the picture in the US was of an industry which was outmoded, with too many plants operating (there were 50 integrated mills whereas 25 were probably sufficient). In 1975 the steel industry began to enter a crisis phase accompanied by slowdowns, closures and a wave of disinvestment. Between 1977 and 1981 some 14 million tons of mill capacity were closed down, although this was in part counteracted by the erection of approximately 5 million tons of mini-mill capacity (Hogan, 1983, p. 26).

By 1983 some 27 per cent of the capacity of the eight largest firms in the USA representing 130 million tons of capacity, was lost – probably forever. This dramatic reduction in capacity was the result of large multiple-unit steelmakers closing down obsolete or valorized plant. This generally took place in older traditional steel-making towns such as Youngstown, where it was considered the 'brownfield' plant was too expensive to replace (Lynd, 1983). With such massive attrition, the steel industry ceased to be an employment growth sector. As a result some traumatic plant and community ruptures and job losses occurred: between 1973 and 1984 the steel workforce was reduced from 500,000 to less than 300,000. While some steel company executives congratulated themselves on the new slim look of the industry, the job losses have meant significant alterations in regional employment. Many workers will never be re-employed in the same steel mill, many never again in steel and some will never work again at all. There are few jobs available for steel workers in the short term in new expanding sectors such as services or light assembly and manufacturing.

Why did this critical path evolve in the US? A number of reasons may be considered. A critical and absolute fall in profit occurred during the crisis phase. There were increases in costs of steel scrap, iron ore and energy, accompanied by a slowing of reinvestment in new technology, in new capacity and in new mills. In comparison with the major competition from Japan there was a production cost crisis which major firms could not overcome. Most steel-making equipment and plant in the US was installed either before, or during, the 1960s and did not embrace a number of new and important technological items which were standard equipment in Japan and in Canada. The US industry was dominated by open hearth technology, a lower percentage of basic oxygen furnaces and no significant continuous casting facilities. Indeed, the installation of such features and of general modernization programmes was delayed and eventually became too costly. The whole production system therefore, with the exception of mini-mills erected in the 1960s (never more than 15–20 per cent of total production) was less productive, but still too expensive to cast aside and embrace a wholesale adoption of Japanese methods and machinery. By 1980, some 80–85 per cent of Japanese steel was produced in integrated mills incorporating oxygen furnaces and more modern versions of that type of technology: only 65 per cent was produced in a similar fashion in the United States. The result was that by *c*.1982 the average costs of production in the United States were still higher than those of Japan. Furthermore the slow rate of adoption of continuous casting technology (which uses 15–20 per cent less labour time) in the US meant that the rate of production and the total numbers of tons per employee were still lower in the United States (Pond, 1985, p. 77).

Such was the extent of the steel crisis in the US that major companies became unprofitable; some $3.5 billion (US) in losses was claimed in 1982 alone, largely as a result of closings and firings. Many steelmakers, realizing that steel was no longer a leading edge, and that a new boom would be unlikely to arise from the industrial wreckage and carnage of the 1980s, restructured investments and switched into alternative areas. For example, Armco invested in oil field equipment and nuclear power plants; National Steel invested in aluminium, metal processing, building and construction and containers; US Steel switched to mining, cement, petrochemicals, natural gas, oil and, reputedly, Disneyland.

In 1985 the industry as a whole was continuing to attempt to improve its competitive position and its rate of profit. A number of continuous casting plants were installed and the extension of new investments into new greenfield sites in mini- and midi-mills left the brownfield sites as relics of a bygone phase of capital accumulation. For the most part the steel companies survived with a new 'leaner' look, but a number of traditional steel towns have been significantly altered. Many steel plants have no unions; contracts and negotiations have been replaced by no-strike agreements. The United Steelworkers of America (USWA) has already made big concessions, which in some cases will be given back over a period of years. However, it is highly unlikely that any further major concessions on an industry-wide basis will be made in the near future, although the creation of non-union areas is likely to continue, especially where new mini- and midi-mills are located (Hogan, 1983, p. 26).

LONG TERM RESTRUCTURING IN CANADA

What is clear and significant in Canada is that both labour and capital were not subjected to the identical and long-term crisis pressures that have been exerted on the USA, European or Japanese steel industries. Canadian steel producers were not forced into the massive overproduction and restructuring crises which devastated so much of the international steel production system. While it would be easy to dismiss the Canadian crises as less significant by comparison, they are nevertheless an integral part of the dynamics of the system. Unlike the US, Canadian public policy has not been to set quotas on steel imports. Instead, steel companies have lobbied the federal government to set tariffs and to allow transportation costs to operate as a discouragement to foreign steel firms wishing to enter the Canadian market (Cheung, Krinsky and Lynn, 1985, p. 9).

The state has also controlled imports through anti-dumping laws to which Canadian producers appeal from time to time. The mean level of tariff protection in 1984–5 was approximately 7.3 per cent, which was around 1 per cent higher than mean rates in the US and in most European countries. These strategies, however, did not deter the entry into Canada of foreign steel originally intended for the US during the overproduction period leading up to the 1981–4 crisis. In the past the national industry traditionally relied upon exclusionary strategies to maintaion high levels of local production and to permit imports only during periods of peak demand. This was an important element in the maintenance of a reasonable control in what was generally a stable indigenous division of labour.

While Canadian steel companies have traditionally relied upon tariffs to protect some of their capacity, they have never had the level of protection which has existed in Great Britain, Japan or the US. Indeed Canadian steel makers have a somewhat conservative reputation in the sense that they did not, or could not, follow the massive expansions and chase for international competitive markets which marked so much of the post-1960s and mid-1970s activities of steel makers elsewhere. This lack of participation in the international steel orgy of overproduction and competition is probably what prevented the extent or depth of industrial crises from devastating Canadian steel in the late 1970s and early 1980s (Warrian, 1984).

Between 1945 and about 1975 Canadian steel companies operated on the basis of a generally vigorous period of expansion with fluctuating but reasonably high rates of profit. There was a ready availability of investment capital (most operations were self financed) which permitted the development of a specific strategy for the design and construction of plant. This involved constructing a mill with a lifespan of about thirty years, which would run from one scheduled maintenance period to another with minimal downtime for repairs and with a built-in margin for expansion within brownfield plants, allowing for appropriate modernization of old facilities through new technology (Allan, 1984, p. 20). The timing was generally dictated by the availability of capital and by means of a fairly stable local and protected market. Designing plant and machines for a prolonged period fitted in with the longer life-span of product lines. In the crisis of the 1980s, however, with low economic growth, low steel demand, small operating margins and higher costs of financing, steel mills were required to change more frequently in terms of products, machinery and design. The turnover time of investment and the period of valorization of capital, was thus much shorter, whilst the capacity for older steel mills to remain competitive became significantly lower. The price of restructuring in a technical sense was also much higher; new forms of management, computer networks and research and development were required to become part of the restructuring strategy of capital. What was created by capital was '...an environment in which secondary producers (have) become more consumer oriented and basic producers more customer oriented than at any time in their history' (Allan, 1984, p. 20). What also happened at the same time (especially between 1982 and 1985), was a 25 per cent loss in the total number of jobs in the whole Canadian steel sector – a sad end to a traumatic period of economic restructuring and technological change.

CONCLUSION

Restructuring involves changes in the order and timing of investments in technology in plant and in the methods of production. The impacts of such changes can be traumatic in both economic and social terms to employees as well as to capital. The process embraces both long- and short-term changes and is not limited to periods of economic downswing or to the immediate survival strategies of firms in adverse competitive situations. The corollary of 'restructuring' is, therefore, 'structuring', which implies the positive sequencing of events, sometimes in critical situations, but whose long-term objective is to obtain expanded production, effect the reproduction of fixed and variable capital and to continue capital accumulation.

This broad process of restructuring occurs with different levels of intensity over time. There is a reasonably regular pattern to the technological and investment sequences which occur in a winding-down phase of restructuring: slowing down of the production process, intensification of work processes, followed by closures and layoffs or dismissals. The winding-up phase is virtually the opposite. However, these simple sequences do not fully explain the reasons for complex temporal and technical behaviour of an industry such as steel making as it moves through phases of expansion, contraction, or over-production. Usually the pattern has followed business cycle fluctuations with an appropriate lead or lag time depending upon whether one is observing the upstream or downstream linkages. While there is a degree of equanimity between these events so that they can be regarded as normal cycles, the explanation of the events which result in traumatic crises must be sought elsewhere.

In the case of the 1980s crisis, the disequilibrium phase arose only after a long series of events which were imposed upon the 'normal' cycle. These may be termed 'trigger events' and in the present case they arose over a long period. They culminated in the oil/energy crisis; overproduction by most world producers; product cycle crises (namely through inputs to user industries such as automobile manufacture); intercapitalist conflict and higher levels of international competition. The result was a reduction in the absolute level and the rate of profit, to the extent that many major steelmakers delayed investments in plant expansions and in modernizations, switched investments of valorized capital to other industries and sectors and laid off and fired their employees.

Furthermore, it is necessary to explain why the long term (twenty year) expansion of the global steel industry resulted in a crisis, when, in general terms, the events of the 1960s and 1970s pointed to a potentially longer period of expansion and 'cooperative' competition at the international level. Essentially the argument must focus on the dominance of competition and on the absence of cooperation in a global formation. The international division of labour for many production systems is, in fact, a series of national divisions. In the case of steelmaking these are often generously endowed by governments and jealously guarded by the state, except when their commodities reach the international market. Add to this the burgeoning industrial progeny of NICs, many of whom seek badly needed foreign incomes, combined with the aging technology and machinery of the steel industries of western Europe and North America, all of which have a tendency toward overproduction of (semi-finished) steel, and a classic state of disequilibrium and crisis results.

Part of the explanation of crises and of the need for restructuring lay in the international and national linkages of steelmakers. However evidence must also be sought within the production process and the social relations of production. The most obvious reaction within the restructuring of labour comes from layoffs and dismissals with the subsequent community ruptures, unemployment and personal, social and cultural dislocation. While due attention is not paid to this problem here, the evidence is present in most older mill towns of Europe and North America. Such is the intensity of the restructuring of social organizations at the local level, that class conflict in the workplace becomes focused entirely on the attempts to retain jobs and to maintain the integrity of the local operations. Attempts are made to bargain with the industry and with the State, to ensure local jobs by means of financial aid in the first instance. Strikes may take place to pre-empt closure and attempts have been made to create collectives or cooperatives to purchase plants. In communities dominated by a single industry, closures have been accompanied by the reorientation of class positions and class places among local businesses and workers.

In the long run there is a tendency to substitute capital for labour (workers by machines). While this is presently increasing in many industries, there is too often a tendency to view all labour as being continually substituted or ultimately substitutable. The great periods of technical change in steel making, at least in Canada and the US, took place between 1900 and about 1920. At that time, production machines and cranes replaced a high level of labour input. While the technical component subsequently continued to rise, to the point where steelworkers became skilled machine minders, there has never been a repeat of such a significant event in steelmaking in terms of labour replacements. In modern steel plants there are fewer and fewer workers, measured in terms of worker hours per ton, largely as result of the expansion of steel making technology and of the continual modernization of machines and equipment (continuous casting technology being the most recent and notable form of this).

As well as substituting technology for labour power, there is also a constant drive in the restructuring process toward substituting new technology for old. All of this may displace workers, but the intention is to maintain a pattern of constant reproduction of part or whole of the fixed capital. This is because in a capital-intensive industry (Department I)[1] such as steel making, the machinery is relatively durable but it still requires maintenance which is expensive. Furthermore, its efficiency must be updated in order to keep it competitive and in order to maintain it in a condition whereby it can be adapted to produce different types and grades of commodities. The pace and direction of technological change in steelmaking is therefore of great importance.

Under the rubric of crisis theory it is possible to see that the balance of technology, capital costs, machine repairs and labour costs is increasingly exposed to factors of disequilibrium. Indeed, the capital costs of avoiding disequilibrium, and hence of entering into a crisis-restructuring event, are very high, because of the need to link machine repair costs with machine replacements, some of which may not have become valorized. In addition, the costs of new technology to improve the efficiency of the existing machines and of the technology required for such major changes as continuous casting, have become extremely high (hence the move toward mini- and midi-mills in North America designed to avert major outlays required for large-scale fully integrated mills). When such high-cost technology barriers coincide with external events, such as energy costs or overproduction, the long term impact on parts of the industry is traumatic.

The immediate source of free funds to apply to capital circulation becomes an additional problem; the high cost of financing, especially in a very competitive field in a recession, becomes a barrier to capital. In the case of a number of national steel industries, excluding for the most part Canada and the US, the State is called upon to step in either to provide devalorized assistance capital, or to create regulations to protect existing plant and investments, or to create and shape geographical market structures.

The geographical outcome of restructuring in North America is evident in the uneven regional impact of investments and disinvestments in the steel industry. The effects of layoffs are seen in mill towns in both Canada and the United States (Lynd, 1982); community ruptures, social network changes and fluctuations in local businesses are clearly recognizable. However, the longer-term effects of permanent closures and of unemployment are yet to be seen and measured. The immediate trauma of closures and of the removal of the primary local economic base has been documented elsewhere, but the long-term effects of regional shifts in industry and employment may not be known until the current period of high level activity in restructuring, and the juggling of sunset and sunrise industries, either stabilizes or is complete.

NOTES

1 Department I: branches of capitalist production producing means of production (raw materials, energy, machinery and tools, buildings). Department II: branches of capitalist production producing means of consumption (consumer goals), which reconstitute the labour force of the direct producers and contribute to the livelihood of the capitalists and their dependants (Mandel E. (1972) *Late Capitalism*, New Left Books, London, p. 593).

REFERENCES

Aglietta, M. (1979) *A Theory of Capitalist Regulation: The U.S. Experience*, New Left Books, London.
Allan, J. D. (1984) Iron and Steel technology and the market place, *Iron and Steel Engineer*, December, 20–22.
Aylen, J. (1983a) Technology looks for new directions, *Iron and Steel International*, February, 40–43.
(1983b) Are we doing enough, *Iron and Steel International*, August, 129–132.
Barnett, D. F. and Schorsch, L. (1983) *Steel: Upheaval in a Basic Industry*, Ballinger, New York .

Bradbury, J. H. (1984) The impact of industrial cycles in the mining sector: the case of the Quebec–Labrador region in Canada, *The International Journal of Urban and Regional Research*, 8, 311–31.
(1985) Regional and industrial restructuring processes in the new international division of labour, *Progress in Human Geography* 9, 38–63.

Bradbury, J. H. and Martin, I. St. (1983) Winding-down in a Quebec mining town: a case study of Schefferville, *The Canadian Geographer*, 128–44.

Canada Department of Employment and Immigration. *Layoffs and Terminations in the Steel and Steel Products Industry, 1982–85*, Ottawa, Canada.

Canada Statistics *Iron and Steel Mills*, Cat No. 41–203. Annual.

Cheung, C. S. Krinsky, I. and Lynn, B. (1985) The Canadian steel industry: current state and future prospects, *Canadian Banker* 92, April, 7–13.

Clark, G. L. and Dear, M. (1984) *State Apparatus*, Allen and Unwin, Boston.

Harvey, D. (1982) *The Limits to Capital*, Basil Blackwell, Oxford.

Hogan, W. T. (S J) (1983) *World Steel in the 1980's: a Case for Survival*, Heath, Ontario.

Hudson, R. and Sadler, D. (1983) Region, class, and the politics of steel closures in the European Community, *Environment and Planning D* 1, 405–428.

Iron Age (1984a) January 16, 23.

Iron Age (1984b) May 7, 53.

Iron and Steel International (April 1982).

Iron and Steel International (1983a) April, 75.

Iron and Steel International (1983b) October, 183.

Iron and Steel International (1983c) December, 215.

Iron and Steel International (1984) August, 131.

Kawahito, K. (1982) Japanese steel in the American market: conflict and causes, *Iron and Steel International* April, 94.

Lynd, S. (1983) *The Fight Against Shutdowns: Youngstown's Steel Mill Closings*, Sidejack Books, California.

Mandel, E. (1972) *Late Capitalism* New Left Books, London, 593.

Massey, D. (1978) Capital and locational change: the U.K. electical engineering and electronics industries *The Review of Radical Political Economics*, **1039–54**.
(1983) Industrial restructuring as class restructuring: production decentralization and local uniqueness, *Regional Studies* 17, 73–89.
(1984) *Spatial Divisions of Labour*, Macmillan, London.

Massey, D. and Meegan, R. (1979) The geography of industrial reorganization, *Progress in Planning* 10, 155–237.
(1982) *The Anatomy of Job Loss: the How, Why and Where of Employment Decline*, Methuen, London.

Moulaert, F. and Salinas, P. W. (Eds.) (1983) *Regional Analysis and the New International Division of Labour: Application of a Political Economy Approach*, Kluwer–Nijhoff, Boston.

Morgan, K. (1983) Restructuring steel: the crises of labour and locality in Britain, *International Journal of Urban and Regional Research* 7, 175–201.

OECD *The Iron and Steel Industry* Annual Reports.

Pond, J. B. (1985) Iron Age's 1984 top 50 World Steel Producers *Iron Age* May 3, 77–82.

Smith, N. (1984) *Uneven Development*, Basil Blackwell, Oxford.

Warrian, P. (1984) Saving the steel industry, *Our Times*, 16–20.

PART III

Technical Change and Spatial Policy

9

Technical change and the decentralized corporation in the electronics industry: regional policy implications

D. R. Charles

The implications of 'external control', whether viewed at a national (Britton, 1980; Economic Council of Canada, 1983) or a regional (Watts, 1981) scale, have been a major focus of research and policy interest during the 1970s and 1980s. In the case of the UK assisted areas, the proportion of employment in externally owned plants ranges between 40 per cent and 80 per cent (Dicken and Lloyd, 1978; Smith, 1979) and even in non-assisted areas, city regions such as Southampton or Bristol may have very low levels of indigenous control of a similar order to the peripheral regions (Mason, 1981; Bassett, 1984). 'External' can, of course, mean that ultimate ownership resides in an adjoining region or on the other side of the world. However, the significance of foreign direct investment (FDI) in UK manufacturing seems to be higher in 'technology-intensive' sectors such as chemicals, motor vehicles and aerospace. This trend is especially apparent in electronics, both at a European and UK scale. Table 9.1 emphasizes the significance of US-based electronics firms in Europe while the indigenous firms have their headquarters in a small number of core regions. The implications of this are exemplified by Scotland where, in 1979, 47 per cent of employment in electronics was in foreign-owned plants with a further 43 per cent in English-controlled establishments (Booz, Allen and Hamilton, 1980). Thus in 'technology-intensive' sectors in general and electronics in particular, many manufacturing establishments in peripheral regions of the UK are under external control.

The degree of technological autonomy is one dimension of the branch-plant

Table 9.1 HQ location of 100 leading electronics firms in Europe

Nation	No. of firms
UK	13
France	9
West Germany	15
Netherlands	2
Belgium	1
Italy	4
Denmark	1
Switzerland	7
Sweden	3
Norway	1
Finland	2
USA	28
Japan	13
Hong Kong	1

Source: Mackintosh, Annual Reports

syndrome. Given the spatial concentration of research activities within the 'technology-intensive' sectors of the economy (Howells, 1984), then it is clear that there is a danger of the peripheral plants lacking the ability to react to technological change and being subject to uncertainty in the wake of decisions made elsewhere. Research and development (R&D) and innovation are frequently identified as a major component of corporate success (Pavitt, 1980; Old, 1982; Mansfield, 1981; Economic Council of Canada, 1983; Abernathy et. al., 1981) and it is logical to presume that in many cases this may also be true for establishments. Plants which incorporate only basic production technology, lack R&D and are also peripheral within a corporate structure seem most vulnerable to closure (Massey and Meegan, 1979). The potential for secondary multiplier effects at the regional scale will also tend to be reduced in such plants. It is, therefore, important to have some awareness of the extent to which R&D is carried out in branch plants. The next section of the chapter examines the evidence for this and enlarges on the policy implications before exploring one possible strategy for change. These issues are then explored in the context of the electronics industry.

R&D IN BRANCH PLANTS

The presence of R&D at a branch plant is one of several indicators of its degree of autonomy. These indicators include other functions such as purchasing, marketing, production control and service/maintenance (Hoare,

1978). Wood (1978) suggests that the need for the geographical association of such functions with production, and hence the extent of local autonomy depends upon certain characteristics of a firm's activities. These include the manufacture of unstandardized, innovative products using small-scale, batch processes, reliance upon labour-intensive methods based upon specialized skills, and involvement in uncertain, competitive and variable markets. All of these characteristics are likely to be associated with industries undergoing high rates of technical change and certain sectors are, therefore, more likely to provide conditions favourable to the establishment of autonomous operations.

Unfortunately, the typical branch plant is engaged in the manufacture of mature standardized products and there is much evidence indicating the low level of R&D in such plants. Several studies draw attention to the spatial concentration of major R&D facilities both within the UK and in other nations (Buswell and Lewis, 1970; Howells, 1984; Malecki, 1979; Lipietz, 1980) and this can be shown to be the consequence of a number of processes. The reduction in indigenous control by merger activity has been a major cause of spatial rationalization since the early years of the twentieth century. The creation of ICI in the 1920s, and of major electrical companies like GEC and AEI in the same period began the trend towards the establishment of major centralized laboratories, especially around London, with a reduction in advanced development work in previously autonomous provincial locations. (Hannah, 1976; Jones and Marriot, 1970). This process has continued in more recent times with rationalization after acquisition and the imposition of corporate hierarchies onto a range of plants often involving the downgrading or even closure of plants in peripheral areas (Leigh and North, 1978; Massey and Meegan, 1979).

The growth of greenfield investment, particularly by US firms in the wake of regional policy, has also encuraged polarization. Hood and Young (1977) have estimated that in Scotland, 55 per cent of US-owned branches and 64 per cent of European branch plants have no R&D whatsoever. This latter situation is mirrored in Canada and many European nations, although one suspects a different story with European plants in the US, many of which are aiming to exploit technological comparative advantages.

In research conducted at the Centre for Urban and Regional Development Studies, Thwaites, Oakey and Nash (1981) found that the plants least likely to have R&D were single-site independent firms in peripheral regions and branches of multisite corporations. Interestingly, the incidence of R&D in mutliplant enterprise establishments in the development areas did not differ greatly from that in other regions. This suggests that the problem of the development regions is not so much one of differences in the properties of branches between the regions, but in the different regional mix of types of establishments, particularly with regard to the presence of higher function decision-making units. If this is true there should be concern about the possibility of branches being converted to complete business units with greater positive multiplier effects rather than remaining as limited function establishments.

The regional benefits from an increased level of autonomy through on-site R&D can be summarized as follows:

1   On-site R&D may lead to improved products and processes allowing the rejuvenation of those plants based on mature technology, and growth and diversification into new markets and product areas. This inevitably benefits both the corporation and the region through increased profits and backward linkages.

2   The higher functions required to enable R&D, innovation and resource allocation to take place necessitate professional and technical staff not typically associated with branch plants, so providing a wide range of employment opportunities in the plant and therefore the region.

3   These higher functions also imply demands upon the local producer services and subcontractors as local autonomy must increase to enable the R&D input to be effective, allowing the plant to plan some local contracts instead of relying upon services arranged through intra-firm control hierarchies.

4   Increased proportion of professional employment and local producer services within the region improve the environment for new firm formation, innovation, adoption and then further investment.

These mechanisms can be demonstrated from various pieces of evidence, although more research is needed to understand some of the mechanisms involved. Haug, Hood and Young (1983) indicate some of the advantages of on-site R&D in US electronics plants in Scotland. Greater R&D intensity appears to result in a more export-oriented nature and a higher potential as a primary source of components to companies located outside Europe. It results in greater demands for graduate employees both in numbers and quality and may encourage in-flow of these from outside the region. Finally it seems to benefit technology transfer, speed up production learning curves and increase the numbers of new projects based in the affiliates.

This background has led to arguments for policies focused upon increasing the provision of R&D facilities in all types of establishment. Specific schemes to achieve this can be identified both in the UK and in Canada, although in the case of the latter these aims have been much more explicitly stated than in almost any other nation. Thus there has been the Canadian range of tax incentives, and grants such as the Industrial Research Assistance Programme. A bewildering array of technical support schemes and industry-specific programmes can be added to a policy environment where governments have been convinced that 'the application of new technology in Canadian industry can improve industrial competitiveness and productivity and thus can increase growth prospects and employment opportunities' (Lalonde, 1983). In the UK, similar intentions lie behind the Support for Innovation scheme, a plethora of sectoral programmes and the Office and Service Industries Scheme (OSIS). OSIS in particular is geared towards encouraging the movement of service functions to assisted areas and could be said to be encouraging decentralization of control in multisite firms, through the decentralization of office or research functions.

However the evaluation of such policies has so far been fairly critical. In Canada the main criticisms fall into two camps; some argue the futility of fighting an international structure determined by comparative advantage (Globerman, 1978), others that policies have been too 'research' oriented and have neglected both the problems of putting advanced products into the market, and the ability of branch plants to develop products in the face of corporate control structures. The problem of product development and launch is especially difficult for firms in Canada where costs are high, domestic markets are small and overseas markets are generally protected and dominated by larger competitors. In Britain these policies have largely seemed to be ineffectual at altering branch-plant economies, being too restricted financially to affect the decisions of major actors who feel far more constrained by shortages of qualified manpower, and who can usually obtain some support for R&D in core regions from other schemes.

The failure of policies designed to encourage the dispersal of R&D within multiplant enterprises ultimately reflects corporate resistance to the dissemination of the knowledge upon which firm-specific advantages are based. Such resistance leads to the centralization, in both an organizational and geographical sense, of this knowledge. Centralization imposes costs of control and coordination and the degree of autonomy will vary in proportion to these costs (Rugman, 1981). The prospects may be especially poor in the case of foreign-owned branches as the technological capacity may be retained in the home country, although the added distance, cultural, institutional and tariff barriers may encourage some degree of autonomy. Rugman however argues the futility of non-discriminatory R&D incentives, because the possible take-up by companies may give results fundamentally opposed to the aims of policy as information is procured and diffused back to the core of the firm rather than to the operating milieu of the branch.

WORLD PRODUCT MANDATES AND THE DECENTRALIZED
CORPORATION

One possible solution to the conflict between the corporate tendency to centralize R&D and the policy objectives of regional development agencies is the world product mandate which allows the multinational to use its international marketing power to sell a specific product line produced by a single subsidiary on a worldwide basis (Science Council of Canada, 1980). This policy option is especially interesting because it may accord with future trends in the structure of business organizations. The experience of a small number of firms in Canada which have adopted the world product mandate system has revealed benefits to both the multinational and the host economy. From a corporate perspective, it allows the maximization of scale economies through increased specialization and production for export markets. For the Canadian economy, it is suggested that internal linkages can be improved and that R&D facilities may be introduced as a necessity for the effective exploitation of the mandate. Corresponding benefits may be experienced at the regional scale, although the effects of export balances will be subordinate to the impact of increased local functional autonomy. However, in view of the negative interpretations of Rugman (1981) noted earlier, is optimism regarding this possibility justified and can effective policies be oriented around this concept?

This type of organization has so far been very much an exception to the norm with foreign-owned branches in particular displaying very low R&D intensity levels (Hood and Young, 1977; Britton, 1984). Also, within the indigenous corporate sector the spatial concentration of R&D units (Buswell and Lewis, 1970; Howells, 1984) is a reflection of their centralization within the organizational framework of the firm. Malecki (1980) draws parallels between R&D structure and corporate structure; thus in centralized firms R&D is concentrated in central facilities rather than the operating divisions. Nevertheless, there are also circumstances in which small development units may be established in association with branch plants. A number of processes are increasing the opportunities for choice in this organizational/locational pattern, although as yet there is little firm evidence on the eventual outcome.

At the international scale many MNC's are integrating their branch plants into global strategies, with the replacement of the satellite business – a scaled-down copy of the overseas parent – by integrated plants making the most effective use of capital. This increased specialization of production has in the past often worked against the autonomy of the branch, but an increasing global sphere of operations and an unwillingness to be reliant upon one nation for development can give opportunities for offshore plants to take the initiative in several product areas and create units to develop their product responsibilities. Associated with this are other processes encouraging the development of a degree of externalization and subcontract work with 'Just-in-time' delivery schedules. The requirements of this organizational innovation are such that the control by the main contractor is best achieved through proximity to suppliers (see chapter 7). To be effective this system depends on increased control by the plant over decisions relating to suppliers, such as the specification of technical requirements.

Despite the growth of these globally oriented plants, another technologically driven trend is towards flexibility and destandardization of product. The adoption of flexible manufacturing systems enables rapid adjustment of lines to model changes and greater product variety, but without the learning curve inefficiencies of more traditional production methods. However, the response to consumer demand for destandardization requires greater flexibility and imagination on the part of the workforce (Toffler, 1985). As a result, worker participation, quality circles and matrix management are introduced, breaking away from traditional hierarchical production systems and creating new opportunities for decentralized R&D. This trend is accentuated when a greater proportion of development work involves reducing costs through altering designs to fit the manufacturing process. Proximity to manufacturing then exerts a greater pull than the creative isolation of a centralized R&D unit. Information technology is another potentially decentralizing influence, although its impact upon corporate structures is by no means clear (Child, 1984).

The overall effect of these developments on R&D dispersal and branch-plant autonomy is uncertain. Under what conditions will decentralization of R&D take place? What implications will it have for regional economies? Will new decentralized corporations grant product mandates to their branches and will this occur in all types of region? The following case study of the electronics industry considers these questions.

INDUSTRY STRUCTURE AND TECHNICAL CHANGE
IN ELECTRONICS

The electronics industry has been perhaps the most flexible and dynamic of all manufacturing sectors in recent years, and, despite oligopolistic trends in some sectors, it continues to have a high degree of uncertainty. This can be illustrated by referring to four characteristics of the industry;

1 changes in technologies and markets
2 changes in investment levels at a firm and industry scale,
3 changes in managerial awareness and organizational structures, and
4 the resultant market structures and segmentation.

The electronics industry is noted for the pace of technological change, probably the most rapid of any industry when measured in terms of product life spans. Innovation has been propelled by both inventive pressure and market demand, and is now affecting product and process development in many other industries. Public procurement has been a major spur to many sectors involved in defence and data-handling equipment, while the rapid pace of innovation has been fuelled by uniquely generous development grants.

The uncertainty engendered by technology-based competition has created opportunities for small innovative firms dependent upon fast imitation and improvement rather than offensive research-based strategies. The importance of design and technology implementation has enabled engineers to use unique skills in developing competitive products without resort to large capital requirements or large markets. The customization of many product markets has also offered additional avenues for development and, again, innovation has been instrumental in creating differentiation. Thus technological change has allowed the emergence of small-scale operations in innovative and differentiated markets and enabled entry barriers to be overcome at points of radical product change. However, this is not the case for the standardized markets such as office machinery or components.

Technological change has also had profound effects upon levels of investment and hence upon market structure. In the early days of electronics, many of the innovation were relatively cheap to implement. The products were small and final assembly could be done by hand with a low degree of capital intensity – the microcomputer is the latest example where the technology is embodied in the chips and its software and the assembly operation is relatively unskilled. Even in chip production, the early days were not capital intensive compared, for example, with a chemical or pharmaceutical plant, and firms raising a few million dollars from venture capitalists could enter the industry relatively easily.

The position changed substantially after 1975 as a consequence of increased international competition and technological change. The microelectronics industry was transformed as the increasing densities of components on a chip necessitated resolution limits approaching the wavelength of light and the costs of new production equipment for each new scale of integration have consequently risen to levels at which venture capital is not available. The current cost for a typical wafer fabrication plant is between $100m and

$200m and such a plant is unlikely to last five years at current rates of redundancy. Hence total investment for 1984 in a demand-dominated situation was $6bn (*Financial Times*, 1985). Investment has now become the main plank for competition, with the bigger companies hoping to force smaller firms out of the volume market where marginal capital/output ratios are approaching unity. Recent problems with the Inmos and Mostek companies clearly show the success of these policies; with Inmos dropping out of the dynamic RAM market, and Mostek being closed down by its parent. High investment levels are not only necessary in component manufacture. Technology requirements in the computing and telecommunications industries have made it impossible for smaller firms to initiate substantially new models. The development costs of the System X exchange were around £300m, a figure which looks small when set against IBM's R&D expenditure of $15.5 billion between 1980 and 1984. The UK government's Alvey programme, to develop expertise in fifth generation computer technology, with an expenditure of £350 million is not only insignificant by comparison, but its collaborative nature clearly illustrates the difficulties facing relatively small UK-based firms competing in the computer industry at a worldwide level. This need to compete in a worldwide market also inevitably results in increases in marketing budgets, with similar implications for cooperation and merger.

There have also been considerable changes in managerial philosophy over the last twenty years as the demands of technology and competition have undermined the mechanistic hierarchies of the 1960s. A major concern has been of managing for change, and hence organic structures as demonstrated by Burns and Stalker (1966) have become more popular. Typical developments include corporate image-building, market orientation, flat organizations, small teams, competitive projects, and lucrative reward systems. The pressure of the Japanese in many markets has been a further stimulant and many management characteristics presumed to be factors in the success of such companies are becoming widely adopted, typically with the Theory Z concept in the US (Ouchi, 1981). A number of companies now try to reproduce the paternalism of the Japanese with increased responsibility and consultation.

The small operating unit has been an important part of many of these changes. Many small firms emerging in new product markets have been able to grow without sacrificing their small-scale informality since the technology does not need giant operating facilities. Additionally, larger firms have acquired smaller units with entrepreneurial records and have been disappointed by attempts to reorganize them, often losing the key innovators once the small unit is broken up. Evidence also exists for the success of small venture operations in large firms where innovation is encouraged by the absence of other responsibilities and by strong project identification. The academic background of many electronics innovators resulted in a set of values antipathetical to traditional industrial hierarchial management, and the informal management style of the new small firm has thus become seen as an industry norm.

The pressures just described have led to a characteristic industrial structure. Three clearly definable groups can be illustrated, each with specific advantages in particular product markets.

1 The familiar global corporation or oligopolistic sector has come to dominate a large proportion of the industry. This is especially true of those product areas with high R&D investment and marketing requirements, where international operations are essential to achieving viable sales levels. This group is primarily composed of American and Japanese firms with a small number of European companies, examples being IBM, AT&T, Matsushita, Phillips and Siemens. Data processing, telecommunications, advanced components and consumer products are the major sectors where size is usually essential and in these sectors concentration is very pronounced. Thus 80 per cent of worldwide telecommunications equipment sales are made by just twelve firms (OECD, 1983) and in electronic data processing, IBM alone has 56 per cent of current world sales with 70 per cent of all installed mainframes (Locksley, 1981). After 1975 this oligopolistic trend was encouraged by the merger of a number of firms to create information-based companies such as IBM's takeover of Rolm and MCI, and AT&T's moves into computing technology. The spread of the giants into smaller economies is effectively squeezing out the smaller national firms, and IBM has become the dominant local producer in most advanced economies.

2 Lying below the global corporations are a set of companies operating in international markets, but primarily based on one home country and with limited oligopolistic muscle at international level. Typically they may have less than 5 per cent world market, a turnover of less than £2bn, but with a diversified multinational operating capacity. Britain's large electronics firms, with the possible exception of GEC, fit into this bracket, with STC, Racal, Plessey, Ferranti and Thorn-EMI being the principal examples. Often forced out of, or denied entry to, the oligopolist markets such as consumer electronics and volume semiconductors, they are usually able to use firm-specific advantages to exploit specialist, niche or protected markets to build

p their turnover where economies of scale can be outweighed by
product experience and differentiation or preferential public procurement. This is particularly true for defence-based markets such as
radar, radio, security and small computer systems, with components
other than volume semiconductors also included. A number of smaller
telecommunications producers also fall into this bracket, aiming at
specialized or home markets, although these are increasingly under
pressure from rising costs. Undoubtedly there is the possibility of
growth by individual firms to become global in scope based upon
the early development of new technologies, but there is pressure for
merger again as in the late 1960 to allow these companies to gain
the financial basis for competing internationally. These are the firms
that are perhaps most valuable in providing a national technological
capacity and they have been the recipients of much government aid
for R&D. They therefore have an influential role in the dispersal
of R&D at national scales.

3 Finally there is the small-firm sector composed of a wide range of
types and sizes of firms engaged in three main product areas; the
exploitation of newly emerging markets using standard component
technology, the tied production of components and subassemblies
and the production of bespoke and niche products. The survival of
this sector will be partially determined by the attitudes of the larger
firms to subcontracting and the availability of market niches.
Although there are advantages in many situations for smaller
establishments, their success depends upon access to finance and
markets and so upon the sufferance of their larger brethren.

Technological factors seem to be instrumental in an irresistible movement
towards larger enterprises in the electronics industry. This trend is promoted
by increasing scales of investment and technological support and with added
major incentives from the convergence of technologies. The heightened risk
in so many ventures has further encouraged merger, with more competition
in previously protected public contracts and greater technological uncertainty.
Many of the recent problems of the UK electronics sector in the mid-1980s,
indicated by falling share prices, can be attributed to the greater levels of
uncertainty in various subsidiaries and a stock exchange perception that
these companies are not large enough to compete internationally (*Financial
Times*, 1985).

However, this view can be contrasted with the popular media image of
small 'high tech' firms which is often promoted by policy markers. Technological, managerial and marketing pressures are demanding close interaction

between R&D, decision making and production. Even though many of the firms that grew up in the 1960s no longer exist as independents and the image is now being tarnished by the failure of the microcomputer generation of firms, the philosophy of the small establishment lingers on. The questions arise as to whether the larger firms now developing are preserving the adaptability of the smaller units that they acquire, and whether they are adopting this format for their new divisions? Kanter (1983) suggests success emerges from reward systems where seed capital is available for new in-house ventures. Is this a means of maintaining the entrepreneurial zeal within an industry context of high investment requirements? Could this lead on to world product mandates, and what would the consequences be for the branch-plant economies?

ORGANIZATIONAL STRUCTURE IN THE UK ELECTRONICS INDUSTRY

The following evidence for the role of the semi-autonomous unit, or world product mandate in the UK electronics industry, comes from a series of case studies conducted using combinations of interviews, company publications and press reports. This organizational form may arise in four main ways:

1 Entrepreneurial units have developed within large firms in order to exploit new ideas and technologies
2 Small companies have been acquired to gain control of component or related products, or to diversify into new areas, but with the acquired firm being maintained as an autonomous unit after minimal restructuring.
3 Equity investments have been made into small independent firms to gain access to their technology and develop research partnerships, but without taking over complete control, so acting as a venture capitalist.
4 Joint ventures are started between two corporations to exploit specific advantages where synergistic benefits may arise, where the new company is a small innovative operation and control must be limited due to its dual ownership.

1 Entrepreneurial units

The internationalization of the entrepreneurial spin-off is not a recent phenomenon; evidence for such activities in the electrical/electronics industries dates back to at least as far as the 1930s. It is encouraged by the desire of an existing company to retain budding entrepreneurs within the corporate umbrella. This is often encouraged and facilitated by the provision of funds for the development of the new product or process and by the creating of a new company, a wholly owned subsidiary, to manufacture and market the idea.

One outstanding example was Ferranti's Scottish Group, which originated as a temporary wartime armaments works producing gyroscopic gun sights. When the war ended and the contracts expired the decision was taken within the works to resist movement back to Manchester, where most of the management and engineers had come from. Fortunately, three factors were in their favour at this time; the Scottish management under John Toothill was well respected within Ferranti and the industry, the Ferranti family held complete control over the operations and were happy to delegate product/market decisions so long as profits were guaranteed, and the Scottish plant had a highly skilled workforce including the great radar pioneer Sir Robert Watson-Watt. The Edinburgh group therefore developed new radar systems using Minnistry of Defence (MOD) contracts to encourage later diversification into related products and internalized the ability to produce complex radar and avionics products. This attitude, however, has remained Ferranti policy, with no core R&D, small central office staff and development on virtually all sites. The systems nature of much of the work seems to demand high intensity of interaction as the software and hardware must dovetail neatly – the optimal design, critical in military contracts, cannot be left solely to one group to develop. Another factor encouraging contact between R&D and production in the Crewe Toll factory in Edinburgh is the degree of computerization in design, testing, costing, manufacture and stock control; all being integrated to a degree that one recent order was totally redesigned and a new tender submitted within one working day.

Another UK company with a similar attitude to entrepreneurial groups is Racal, which has an informal policy of keeping operating units below 500 employees and, being heavily market oriented, 'a policy of creating autonomous companies to handle individual product areas, each with highly skilled staff possessing the specialized knowledge and ability needed to provide an exceptionally quick response to customer requirements'. Growth is intended to follow two main lines; 'creating additional autonomous companies as the need arises' and acquiring other firms as the opportunity arises. An example of the former policy is the computer-aided design (CAD) subsidiary Redac formed in 1965 and located in Tewkesbury. The role of the individual was very important here again as Eric Wolfendale, a university engineer, was 'head-hunted' by Racal and a research company, Racal Research Ltd, formed to investigate his ideas on CAD in electronics. The establishment was located in Tewkesbury, beside a newly acquired precision engineering company which had both space to expand and a quiet atmosphere for what was advanced R&D at the time. Later, thanks partly to Ministry of Technology funding, the successful products were hived off into Redac Ltd (Racal Electronic Design and Analysis by Computer) and Racal Research Ltd later revived as a microelectronics unit in Reading. Today, the company still operates as an independent unit with no close links to any other Racal company, reporting only to the main board which acts as shareholder. Like other subsidiaries, Redac prepares financial reports with its own board of directors outlining the direction of future investment and predicted performance. This is them submitted to the main board which reviews the budgeting and, unless the performance and budgeting is out of line, approves the expenditure. Competition strategy, however, originates in Redac with its own marketing, R&D and overseas sales offices. This is notable in that the Redac products are not marketed through corporate offices but are under direct control, and there may be several offices for individual Racal subsidiaries in a particular nation, but with no overall company direction.

Another example of this behaviour in a more formal style in a worldwide corporation is Hewlett Packard, where stated objectives to grow from within, relying primarily upon the reinvestment of profits, are coupled with a deep sense of involvement on the part of staff and management. The company is structured into 62 operating divisions within 52 formal divisions, each of which employs a maximum of 2–3000 people. The mechanism for the creation of a new division usually begins with a large unit of around 3000 people with a product range that can conceivably be split into two self-supporting units. These are then separated out on the original site, gradually duplicating management functions to enable independent operation. In some situations they may both develop in situ, in others a move may be required as limits are often placed on the total workforce the firm will allow in particular locations. The more innovative unit with the newest technology and smallest staff will usually be moved, so enabling a new factory to be set up unhampered by old products. The new management team choose a site in combination with corporate staff who supply the real estate and architectural expertise, but the personal preferences of divisional management can still come through as has happened on at least one occasion in the UK. Local autonomy is primarily through product development, as there seems to be increasing corporate control of marketing, but it is normally assumed that each division of 2000 people will have 200 staff in the R&D unit – the optimum size according to Hewlett Packard. World product mandates are standard. For example, the South Queensferry plant near Edinburgh manufactures telecommunications text equipment for worldwide markets and is responsible for business, R&D, manufacturing and marketing.

There have also been entrepreneurial spin-offs by nonelectronics firms such as Mars Ltd, the UK subsidiary of the American food manufacturer. The need for advanced coin-handling mechanisms for sweet dispensers led to an electronic device and a manufacturing subsidiary called Mars Money Systems in the late 1960s. After sustained growth in this sector, the company later decided to develop further markets and expand the electronics business away from the internal markets – the coin-handling equipment goes into non-food areas such as pay telephones. After an investigation of market niches, a gap in small-boat radar systems was discovered, and the Vigil model developed, followed by automatic text equipment, also based upon user experience. The distinctive point about this development is that each of the three businesses has considerable growth potential and could result in a good medium-sized firm, but is attached to an overseas subsidiary and now has branches in the US. Thus it is not impossible for a subsidiary, given a degree of independence, to move into new product areas with a consequent increase in employment in the host country (Mars Electronics employs 600 in the UK) including technological developments new to the host nation.

2 Acquisition of small companies

The electronic components industry has been especially prone to acquisition activity in response to pressures for internationalization and the effect of spiralling investment requirements in the active devices sector. This market in particular has been very unstable over its twenty five year lifespan. Changes have resulted in the decline of the older component firms to be replaced by a bewildering army of small specialist firms, and again a return to a limited number of larger firms with specialist producers in the hands of multinationals with diverse interests.

The forces for internalization are strong; successive periods of boom and bust have created a very unstable market, and those firms with tied supplies have found the benefit of continuity. Also, with rising levels of value added being taken up by chips in a wide range of products, there are substantial financial incentives to establish control. Even in the motor industry, General Motors have felt the necessity to safeguard their access to this technology through acquisition. Despite having its own manufacturing facilities, IBM has also bought into semiconductors, with a 30 per cent stake in Intel. The Intel chips are used in the IBM–PC range and this made up 13 per cent of Intel's sales in 1983 (Management Today, 1985). Similarly, IBM has also been moving into the telecommunications industry with Rolm, a $1 billion digital equipment manufacturer and MCI, a competitor to AT&T on long distance telephone transmissions in the US. However, in both cases it appears unlikely that IBM will be interfering greatly in the running of the firms in the near future, especially in Rolm with its 'laid back' Californian operating style. In spite of its external image, IBM has always decentralized to a considerable extent. The company operates R&D from twenty six laboratories worldwide, in order to gain access to technology from as wide a range of sources as possible. National specialisms are pronounced, with the UK laboratories at Winchester focusing upon software, displays and disks, which feature strongly in the IBM (UK) production range. However, despite this, there is some feeling in the industry that the large size of IBM is a disincentive to radical developments.

On a completely different scale, Racal have used acquisitions as well as internal growth in the development of their network of small businesses, although some of their acquisitions, Chubb, Decca and Milgo, have been considerable businesses. An example of a small operation is MESL, located near Edinburgh and originally a spin-off from the local Ferranti and Decca factories through the efforts of two microwave engineers. The original product range prior to acquisition in 1979 was microwave components, small radar systems and burglar alarms, and the acquisition was on the basis of continued investment to allow expansion. Interestingly, the company was restructured, but yielded three separate businesses. The security business is completely independent and the radar business was merged into Decca Radar, but still operates from the same site with the same product range. Finally, the microwave components business, although also subsumed within Decca, has also gained responsibility for the other factories in the south east and has its own agents and offices on the continent. All three have increased output and employment. As with Redac, each subsidiary prepares its own strategies and budgets, and even though more integrated into the main stream of Racal's business, they must still bid for internal contracts on an open market basis. Each company is a profit centre and so must achieve Racal standards of profitability and growth to avoid attention from the centre.

3 Equity investments

A third organizational trend is for large firms to invest in a minority holding in small technology-based firms in a venture capital sense. The reasons for the strategy may vary from the need to secure supplies against receivership, to gaining access to product technologies or markets, to the need to find a profitable outlet for capital. The most notable case in the UK in the 1980s was the initial 'rescue' of Acorn by Olivetti which was inspired by a desire to take advantage of Acorn's established position in the education field. This is not, however, the first such investment made by Olivetti, which had minority holdings in over 30 companies in Europe, the US and Japan (see table 9.2).

Table 9.2 Significant Olivetti high technology investments

Company	Country	Business	Holding (%)
Docutel	US	Bank automation	46.2
Syntrex	US	Word processing	18.7
File Net	US	Optical disc	11.5
Stratus	US	Fault tolerant computers	10.1
Linear Technology	US	Linear Circuits	5.9
Sphinx	UK	Software	24.0
Dixy	Japan	Flat plasma displays	20.0
TABS	UK	Software	32.0
Editrice	Italy	Software	20.0
IPL Systems	US	n.a.	28.8
Micro Age	US	n.a.	33.4
GTI	France	Software	50.0
VLSI Technology	US	Semiconductors	3.2
Acorn	UK	Mini computers	80.0

Source: Olivetti Annual Report; press cuttings

Ferranti has also made moves in this direction taking up holdings in a small number of companies, partly through a special subsidiary in the US, Ferranti High Technology Inc. This is stated as being expressly for the purpose of 'ensuring that, by investing in promising, new technology start-up companies, we keep abreast of developments complementary to our existing business'. (Ferranti Annual Report, 1983). A secondary, but perhaps more important aim, would be to identify possible acquisitions and investigate their feasibility without committing too many resources. The practice is also widespread in the US with the larger conglomerates taking shares in semiconductor firms in particular, but also in firms in such activities as resource extraction services, CAD, software etc.

4 Joint ventures

The joint venture is becoming a commonly used means of coping with convergent technologies, especially for the medium-sized firms which cannot afford to buy into technology. The industry has a long history of such ventures dating back to the origins of the electric lamp and the Osram works and by GEC and Auergesellschaft before 1914. (Jones and Marriott, 1970)

Although there is much that could be said about multinational operations that are less than 100 per cent owned, the emphasis here is placed on deliberate partnership deals between two corporations seeking mutual benefit, rather than on instances where multinationals are persuaded by governments to allow some degree of local ownership.

The use of the joint venture as a technology-pooling device can be illustrated in several areas of the UK electronics industry. Ferranti has joint ownership ventures with TRW Inc. in subsea production control systems, with Eastman Whipstock in down-hole services for oil and gas wells, with Siemens Ltd in electicity meters, and with GTE Corporation in telecommunications equipment for the liberalized UK market. In all but the Siemens link, the emphasis was on Ferranti gaining access to technology in return for providing access to markets. With 51 per cent holdings in each case, Ferranti has been the originator and dominant partner. The Siemens link was a case of rationalization in the face of overcapacity. Racal has also had joint ventures in artificial intelligence with Norsk Data and in pay television with Oak Industries. Again, in both cases the subsidiary is involved in a technology not previously well represented in the company.

Many other firms have less formal links involving licenses and subcontracts, and the development of collaborative aid projects may lead to more formal joint ownership ventures in the future. Certainly a number of European producers are finding that they are too small to compete in world markets and are looking towards the collaborative project as a means of pooling technologies and resources. It is noticeable that several of the ventures which are emerging are between European firms in oligopolistic markets as well as between European and American firms, for example, Ericsson and Thorn, Bosch and Siemens and Olivetti and AT&T.

POLICY IMPLICATIONS

Overall within the industry, macro- and microeconomic pressures appear to be encouraging both the creation of larger enterprises and the retention of small semi-autonomous operating divisions, with significant implications for local economies. These operating units fulfil many of the requirements of world product mandates; they have better employment characteristics than branches, they may have greater long-term security against closure and have greater linkages with the local economy. However if they are primarily a core economy phenomena, then what hope can be offered to peripheral regions? If the main mechanisms for the development of these units are the acquisition of small businesses and the spin-off of new divisions from existing operations in an organic sense, then the location spread will be confined to the core regions. This is a reflection of differential small-firm formation rates (Storey, 1982) and a low incidence of any but the large, mature-technology branches in the peripheral regions. It should be noted that Scotland cannot in this instance be classed as peripheral as it has a wide range of electronics establishments from branch plants, to R&D intensive operations and small new firms.

If it is accepted that R&D and production increasingly need to co-locate as described above, then the balance of costs and location parameters shift to favour the operation most constrained. As value added for capital and manual labour inputs is quite high in most cases owing to the systems/software context and the use of flexible manufacturing systems, then the production component becomes secondary to the demand for skilled manpower – technologists and technicians. Thus existing operations are locationally constrained by the inability to move the know-how embodied in personnel – few of the firms that have tried this have succeeded, and so new operations, including spin-offs, feel they must give priority to their ability to attract quality staff in a supply deficient labour force. In such circumstances firms have two options: a) to locate in an area with a high density of suitable labour and hope to attract from competitors and new graduates, knowing that they have the security of alternative employers in the same local labour market, or b) to move to an area with a lower intensity of electronics firms and rely upon firm-specific advantages, the attractiveness of the residential environment or new training schemes to gain staff, with a particular problem being the attraction of experienced staff from elsewhere. This latter option is naturally far more difficult and rarely attempted, a reasonably successful exception being Ferranti in Cwmbran. IBM have also used this strategy at times as in Greenock, although later attracting a number of other companies to the same area. It must also be remembered that many areas which have become core locations for the industry were originally new sites with little reputation for electronics developments as in Bracknell, Reading, Havant and Edinburgh.

The issue of labour supply appears to be of prime significance in the initial stages of the industry. The ensuing development of a concentrated labour market due to the modes of growth outlined earlier has created a structure of high and low intensity areas with the former being the preferred location of staff. In a demand surplus situation, the low intensity areas cannot compete for scarce labour resources and so remain unattractive in investment terms. Paradoxically it could be the technician class which is most problematic, being less geographically mobile, but more liable to change firms because they have less project loyalty. The creation of a more dispersed pattern of such labour may be achieved by training for information technology skills in peripheral areas, but this is of little value unless accompanied by investment in manufacturing facilities. Such a favourable combination was achieved in Scotland, initially as a result of an investment-led expansion of labour supply, and subsequently encouraged by the creation of institutional structures to promote an environment favourable to a new indigenous sector. Intervention in this process is not anathemal to high-technology development; an examination of the effects of new town development or defence procurement has demonstrated this (Saxenian, 1985; Lovering, 1985). The growth pattern of the industry clearly shows the importance of new firm formation and research centres, perhaps a feature of any new technology, and the latter of these is certainly open to some influence by government as a major source of research funds.

Decades of decision making have resulted in the concentration of government research facilities in south east England, and recent evaluations of personnel location have focused principally upon lower-order functions. Certainly the creation of centres of excellence in other UK regions could lead to a measure of confidence in the ability to recruit to these areas, and lead growth in selected sectors. Whilst this may no longer be possible with much of electronics-based technology, other sectors are still in an infant stage, and it is possible to envisage a biotechnology strategy for north east England which exploits local strengths in brewing, chemicals and pharmaceuticals.

The old industrial areas may be able to respond to new developments with the rediscovery of their many advantages if cultural facilities, large education labour forces, major education centres and a more cosmopolitan life style. However, this requires effort in industrial educational and environmental development through both private and public investment (see chapter 2). Some cities such as Bristol, Edinburgh, Glasgow and Manchester appear to be going through a kind of renaissance but others are still lagging behind hampered by perceptual problems. Such problems are an important constraints to recruitment in all new industries and policies of environmental improvement may, paradoxically, be of major benefit to industrial revival.

The pressures upon the environment of concentrated development in the south east England are formidable. Nevertheless, many firms and institutions have shown their ability to develop away from the M4 corridor, and it is, perhaps, up to government to demonstrate the possibility of spreading development by dispersing its own activities. If nuclear scientists can be moved to Dounreay, then other scientists can be placed around the old conurbations. Without an R&D base, the labour forces of these areas will be further deprived of qualified staff and the new R&D intensive plants will be difficult to establish and maintain. The branch plant cannot be seen as the solution to uneven development, rather it is a cause, and even in growth industries such as electronics it will continue to be a source of redundancy. Plessey illustrates the pattern well; growth in its southern R&D intensive units, and closure of its older northern telecommunication branches. Re-investment in Liverpool only occurred because of the presence of R&D (Peck and Townsend, 1984), but with few R&D units, the closure of other similar branches will undoubtedly continue.

REFERENCES

Abernathy, W. J. et. al., (1981) The new industrial competition, *Harvard Business Review* September/October, 68–81.

Bassell, K. (1984) Corporate structure and corporate change in a local economy: the case of Bristol, *Environment and Planning A* 16, 879–900.

Booz, Allen and Hamilton (1979) *The Electronics Industry in Scotland: A Proposed Strategy* (a report for the Scottish Development Agency).

Britton, J. N. H. (1980) Industrial dependence and technological underdevelopment: consequences of foreign direct investment, *Regional Studies* 14, 181–199.

Britton, J. N. H. (1984) *Research and Development in the Canadian Economy: Sectoral Ownership, Locational and Policy Issues* (Xerox).

Burns, T. and Stalker, G. M. (1961) *The Management of Innovation*, Tavistock, London.

Buswell, R. J. and Lewis, E. W. (1970) The geographical distribution of industrial research activity in the United Kingdom, *Regional Studies* 4, 297–306.

Child, J. (1984) New technology and development in management organization, *Omega* 12, 211–223.

Dicken, P. and Lloyd, P. E. (1978) Inner metropolitan industrial change, enterprise structure and policy issues: case studies of Manchester and Merseyside, *Regional Studies* 12, 181–97.

Economic Council of Canada (1983) *The Bottom Line: Technology, Trade, and Income Growth*, Supply and Services Canada, Ottawa.

Financial Times (1985), June 27.

Globerman, S. (1978) Canadian science policy and technological sovereignty, *Canadian Public Policy* 4, 35–45.

Hannah, L. (1976) *The Rise of the Corporate Economy*, Methuen, London.

Haug, R., Hood, H. and Young, S. (1983) R&D intensity in the affiliates of US-owned electronics manufacturing in Scotland, *Regional Studies* 17, 383–392.

Hoare, A. G. (1978) Industrial Linkages and the dual economy: the case of Northern Ireland, *Regional Studies* 12, 167–80.

Hood, N. and Young, S. (1977) The long term impact of multinational enterprise on industrial geography: the Scottish case, *Scottish Geographical Magazine* 93, 159–67.

Howells, J. (1984) The location of research and development: some observations and evidence from Britain, *Regional Studies* 18, 13–30.

Jones, R. and Marriot, O. (1970) Anatomy of a Merger: a History of GEC, AEI and English Electric, Jonathan Cape, London.

Kanter, R. M. (1983) *The Change Masters: Innovation for Productivity in the American Corporation*, Simon and Schuster, New York.

Lalonde, M. (1983) *Research and Development Tax Policies: A Paper for Consultation*, Department of Finance, Canada.

Leigh, R. and North, D. (1978) Acquisitions in British industry: implications for regional development. In Hamilton, FEI (Ed.) *Contemporary Industrialisation: Spatial Analysis and Regional Development*, Longman, London, 158–81.

Lipietz, A. (1980) Interregional polarisation and the tertiarisation of society, *Papers of the Regional Science Association* 44, 3–17.

Locksley, G. (1981) *A Study of the Evolution of Concentration in the UK Data Processing Industry with Some International Comparisons*, Commission of the European Communities, Brussels.

Lovering, J. (1985) Regional intervention, defence industries and the structuring of space in Britain: the case of Bristol and South Wales, *Environment and Planning D* 3, 85–107.

Malecki, E. J. (1979) Locational Trends in R&D by large US corporations, 1965–1977, *Economic Grography* 55, 309–323.

Malecki, E. J. (1980) Corporate organisation of R&D and the location of technological activities, *Regional Studies* 14, 219–34.

Management Today (1985) Why Intel married for money, February 1985.

Mansfield, E. (1981) How economists see R&D, *Harvard Business Review*, November/December, 98–106.

Mason, C. M. (1981) Recent trends in manufacturing employment. In Mason, C. C. and Witherick, M. E. (Eds.), *Dimensions of Change in a Growth Area: Southampton since 1960*, Gower, Aldershot.

Massey, D. and Meegan, R. A. (1979) The geography of industrial reorganisation, *Progress in Planning* 10, 155–237.

OECD, (1983) *Telecommunications: Pressures and Policies for Change*, OECD, Paris.

Old, B. S. (1982) Corporate directors should rethink technology, *Harvard Business Review*, June/February, 6–15.

Ouchi, W. (1981) *Theory Z: How American Business can meet the Japanese challenge*, Addison-Wesley, Reading, Mass.

Pavitt, K. (Ed.) (1980) *Technical Innovation and British Economic Performance*, MacMillan, London.

Peck, F. and Townsend, A. (1984) Contrasting experience of recession and spatial restructuring: British Shipbuilders, Plessey and Metal Box, *Regional Studies* 18, 319–338.

Rugman, A. (1981) Research and development by multinational and domestic firms in Canada, *Canadian Public Policy* 7, 604–616.

Saxenian, A. L. (1985) Silicon Valley and Route 128: regional prototypes or historic exceptions? In Castells, M. (Ed.) *High Technology, Space and Society*, Sage, Beverly Hills.

Science Council of Canada (1980) *Multinationals and Industrial Strategy: The Role of World Product Mandates*, Science Council of Canada, Ottawa.

Smith, I. J. (1979) The effect of external takeovers on manufacturing employment change in the Northern Region between 1963 and 1973, *Regional Studies* 13, 421–37.

Storey, D. J. (1982) *Entrepreneurship and the Small Firm*, Croom Helm, London.

Thwaites, A. T. (1978) Technological change, mobile plants and regional development, *Regional Studies* 12, 445–61.

Thwaites, A. T., Oakey, R. P. and Nash, P. (1981) *Industrial Innovation and Regional Development*, Final Report to the Department of the Environment, CURDS, Newcastle upon Tyne.

Toffler, A. (1985) *The Adaptive Corporation*, Gower, Aldershot.

Watts, H. D. (1980) The Location of European direct investment in the United Kingdom, *Tijdschrift voor Economische en Sociale Georgafie* 71, 3–14.

Watts, H. D. (1981) *The Branch Plant Economy – A Study of External Control*, Longman, London.

Wood, P. A. (1978) Industrial organisation, location and planning, *Regional Studies* 12, 143–152.

10
Technology, dependence and regional development: the case of the aerospace industry

D. Todd and J. Simpson

Virtually all governments have embarked on some sort of industrial policy or strategy in order to foster growth and international competitiveness (Adams and Klein, 1983). In large measure, these industrial policies are a response to changing international economic relationships centring on technological change, the spatial restructuring of labour-intensive production, (the offshore flight of capital), and an aversion to technological dependence (Clarke, 1982; Froebel et. al., 1980; Lipietz, 1982; OECD, 1979; Thurow, 1984). These structural pressures on governments have been augmented by immediate pressures arising out of the experience of poor economic performances in most advanced industrial countries (AICs) since the mid-1970s. With techno-logical change taken as the catalyst of economic growth, governments have emphasized the importance of nurturing technologically intensive industries capable of realizing the kind of comparative advantage once enjoyed by the now ailing 'sunset' industries (Magaziner and Reich, 1982; Malecki, 1983; Pavitt, 1980). Consequently there is a consensus that industrial policies need to be focused on specific targets, the so-called 'sunrise' industries or firms.

Ironically, with the waxing of aspatial industrial policy based on the notion of 'picking winners' has come the tendency for formal regional policy to wane. The urgency felt necessary in 'picking winners' scarcely tolerates the ethos of regional policy which in the past frequently involved the bailout of declining industries. Yet, many of the issues germane to a national policy of rearing new industries are also vital to regional development. In truth, regional vitality – just like national competitiveness – is ultimately contingent on the creation of economies freed from the conditions of instability and dependency that are conjured up through the 'branch-plant' syndrome. If such vitality is to occur, regional development planning is indisputably and unavoidably linked to the aggregate industrial policy of promoting appropriate industries. In short, picking winners is conceptually as much a regional as a national issue. The real dispute is not between industrial and regional policy but, instead, concerns the selection of target industries which are suited, at one and the same time, to the needs of both.

The object of this chapter is to explain this synergism between industrial and regional policy – especially as it relates to technology transfer and dependence – with reference to the example of the aerospace industry. This is the quintessential target industry, not only because of its obvious high-tech connotations, but also as a result of its unavoidable dependence on government sponsorship. The latter is occasioned only in part by the industry's 'winner' status: rather, it owes most of its legitimacy to the strategic nature of aerospace and the inordinate research and development (R&D) burden which is an inevitable accompaniment. Unlike the consumer electronics case highlighted in chapter 9, the potential value of aerospace to regional development lies less in the willingness of business organizations to accept regionally based autonomous units and more in its status as a quasi-public sector, beholden to the government for a significant part of its resources and, thereby, amenable to government persuasion as to the siting of its constituent parts. Potential merits aside, this chapter reviews the reality of government attempts to nurture aerospace and speculates on the relevance of those experiences to the integration of high-tech industrial policy with regional policy. To begin with, however, those advantages which render aerospace a supposed 'winner' are identified.

AEROSPACE AS A TARGET INDUSTRY

Governments have generally selected aerospace as a prime candidate for encouragement despite the very real prospect of heavy subsidies over an extended period of time. Indeed, there are a number of persuasive features of the aerospace industry which invariably make it part of national and, occasionally, regional industrial programmes. In the first and most obvious place, it is a major element in the main industrial economies. In the US for example, it generated some $83.5 billion in sales in 1984 and employed more than 1.2 million people (AIAA, 1985). Second, the industry maintains a high exposure to international trade. In 1983, for example, the aerospace balance of trade of $12.7 billion was the best of any US industry (*Fortune*, 9 July 1984). Third, the industry has a high linkage propensity with, for instance, a turbojet engine requiring 10^5 components as compared to the 10^4 used in motor vehicles (MITI, 1982). Fourth, it is a high value-added industry whose value-to-weight ratio exceeds that of many other high-technology products. At the same time, aerospace is a parsimonious user of resources; for example, its total use of resources per worker is less than that in shipbuilding, motor vehicles and steel production. Fifth, as just mentioned, it offers the prospect of massive technology spin-offs from its military aspects; the latter being an area which governments find compelling in any event.

Although treated as a homogeneous entity in industrial policies, aerospace is really constituted from three integrated groups of functions which can be classified as tier 1, 2 and 3 activities. This terminology derives from the definition of the industry as promulgated by the Aerospace Industries Association of America (AIAA). As an assemblage of industries, aerospace focuses on research, development and manufacture of aircraft, missiles and spacecraft in the first place; the production of propulsion, guidance and control systems for these vehicles in the second; and, finally, the procurement of whatever airborne and ground-based equipment is deemed necessary for the test, operation and maintenance of the aircraft, missiles and spacecraft. Tier 1, then, is the design and manufacture of the vehicles themselves (the airframe function), tier 2 embraces the engine and avionics (aviation electronics) makers in the main, whereas tier 3 includes a host of subcomponent manufacturers and repair operations. While a fully fledged aerospace industry finds tier 1 functions indispensable, it also requires the infrastructure provided by the other two tiers if it is not to remain dependent on external sources.

At the national level, most attempts to target aerospace have concentrated on the tier 1 functions, with some afterthought reserved for aeroengine viability. On the whole, the approach in advanced industrial countries (AICs) ultimately revolves around the country's ambitions in airframe and aeroengine manufacture relative to whatever scope is afforded it by the dominant US aerospace industry. In the EEC case, the question is less one of nurturing a free-standing aerospace capability and more one of ensuring that the 'commanding heights' of it (that is, large widebody jetliners, advanced combat aircraft, turbine helicopters and large turbofan engines) do not become a monopoly of the USA. To that end, the Europeans have formed supranational aerospace organizations – Airbus Industrie, Panavia (for the Tornado and, perhaps, European fighter aircraft) and Euromissile are cases in point – which are capable of matching the Americans in programme sophistication and size. The recent Eureka scheme which was initiated by the French as a European civil counterpart to the US Strategic Defence Initiative ('Star Wars') can be viewed as a comparable attempt to compete with the US in avionics and space R&D. For its part, Japan has formally designated aerospace, along with advanced computers and replacement fuel technology, as a cornerstone of industrial progress. It has adopted collaboration on two fronts: as a form of production sharing among aerospace companies in order to upgrade their capabilities in the first place, and as a form of bilateral technology transfer between a foreign (usually American) donor and a domestic recipient in the second. The Japanese Government, through its MITI arm, coordinates the entire process. The prime object here is to consolidate a basic aerospace capability and participate as a full partner (albeit junior) in world-class US commercial aviation programmes. This model is being followed to one degree or another by newly-industrializing countries (NICs) such as Brazil, Israel and Indonesia. Much less ambitious are the target industry approaches in Canada and Australia. In these countries airframe capability has deteriorated as a result of the joint impact of rising R&D costs combined with reduced defence budgets. As well as supporting a residual tier 1 capacity, policies hope to relaunch aerospace on the basis of specialization in niches relatively untouched by the major AICs, and tier 2 activities loom large in these plans (Todd, Simpson and Humble, 1985).

Regional attempts at targeting aerospace fall within this grander rubric. Inevitably, those responsible for furthering regional well being have been influenced by the euphoria for 'picking winners' which so excited macro-level policy making. It is fair to say that much of the interest in aerospace at the regional level took the form of an uncritical acceptance of its purported merits from the national advocates. However, two additional arguments in favour of aerospace have been put forward by regional proponents. First, the industry is apparently 'footloose', rendered so by virtue of its status as an assembly industry fabricating a host of inputs from varied, dispersed sources. Moreover, market access, as conceived by location theory, simply is inapplicable to the industry. In the second place, the industry's continued reliance on design flair and entrepreneurship is indisputable. As Klein (1977) has it, the industry is in permanent need of new interactive firms that are loosely structured with nonhierarchical design teams, so as to maintain the dynamism of technical advance. There are sufficient historical precedents to suggest that the key agent in the process, the designer/entrepreneur, can successfully establish his business virtually anywhere so long as ample backing is forthcoming to overcome initial start-up costs and those entailed in building a market reputation. Under such circumstances, peripheral region sites should not interpose any obstacles in the functioning of neophyte aerospace firms: on the contrary, the willingness of regional authorities to subsidize launch costs should positively boost their chances of success. The veracity of these arguments are tested in the exposition of the Northern Ireland and Manitoba experiences which follow.

While at first sight somewhat incongruous, the selection of the two regions as salutary examples of the problems and potential of targeting aerospace is, on reflection, both logical and obvious. From the administrative aspect, the two regions share a level of autonomy that is denied to many peripheral regions of AICs. Though ostensibly under central government control, Northern Ireland maintains distinct industrial policies and, for obvious political reasons, receives large transfer payments from London. Equally reliant on extensive transfer payments, Manitoba also retains an independent decision-making scope, rendered more telling by the propensity of the electorate to favour interventionist (social democratic) provincial administrations. By the same token, Northern Ireland and Manitoba have been burdened by stagnant economies, and in view of their small populations (1–1.5 million), have been unable to turn to consumer-led growth when their export-oriented industries have run into difficulties. Confronted by massive dislocations in world shipbuilding (see chapter 5), textiles and light engineering, Northern Ireland's increasing dependence on an ill-assorted 'branch-plant' economy has appeared less than convincing. Manitoba's reliance on weak grain and mineral markets has forced that province to contemplate diversification into manufacturing, but of a kind that offers some semblance of permanency (that is, devoid of the usual branch-plant characteristics) in order to counter the endemic instability of commodity markets. Both regions have grasped at high-tech manufacturing as a possible agent of revitalization and aerospace has figured prominently in these endeavours. In the first instance, they were unanimous in advocating tier 1 aerospace activities. Subsequent problems arising from implementation of those activities, however prompted re-appraisals in both cases. This led to strengthening of tier 1 in the one instance but diversification away from it in the other. The manner in which the aerospace option presented itself and the reaction of policy makers to it has differed between the two regions. Nevertheless, their experiences have raised most of the contingencies likely to face other regions with similar aspirations so that general conclusions may be drawn.

NORTHERN IRELAND

State involvement in aerospace in Northern Ireland has conformed to two basic themes. The first is a desire to stabilize existing activity so as to prevent the serious unemployment which would follow from its demise. The second is to encourage establishment of new enterprises in the name of technological progress and qualitative improvements in regional structural mix. This involvement was formalized in 1974 when the major aerospace enterprise, Shorts, was taken over in entirety by the state (although effective state control can be dated to the second World War) and continues in public ownership despite privatization of equivalent interests in Britain itself. Only in the 1980s however, has the state (that is, the Northern Ireland Department of Economic Development, acting on behalf of the UK Government titleholder) acknowledged the economic value of Shorts to its region: its interest hitherto had always stressed the defence aspect. Indeed, the Belfast ariframe enterprise really derives from rearmament in the late 1930s when Harland & Wolff shipbuilders united with Shorts of Rochester, England, to build a bomber and flying-boat plant close to the Queen's Island shipyards. As these facilities were newer and more spacious than the English premises, Shorts concentrated all its aerospace activities at Belfast after 1946 (Corlett, 1981). The first two decades after the second World War saw the firm abandon its indigenous designs in favour of licence production of military aircraft designed by other British firms. Its 'Belfast' military freighter of 1960 was, in fact, a greatly modified redesign of the Bristol Britannia four-engined turboprop airliner. Completion of this programme and a void in defence work obliged Shorts to seek risk-sharing agreements with other firms. These took material form through an arrangement with Fokker of the Netherlands, whereby Shorts would build the wings for the F–28 jetliner and, more prosaically, through a contract with McDonnell Douglas of St Louis for manufacturing the outer-wing panels of F–4 fighters. In fact, subcontracts kept Shorts in business throughout the 1970s, embracing parts for Boeing, Lockheed and British Aerospace airliners as well as nacelles for Rolls Royce turbofans. They sufficed to keep the workforce in regular employment and gave the firm breathing space to entertain the design of new civil light transports. Until the onset of the 1980s, Shorts was really functioning as a tier 2 enterprise; an area of aerospace generally missing from the region with the notable exception of Martin-Baker's ejection seat factory at Langford Lodge (a firm with scarcely any material linkages to Shorts).

Expertise in parts production gave Shorts the resources to develop a family of light transport or commuterliner aircraft. Beginning with the Skyvan, the models 330 and 360 were designed to penetrate world commuterliner markets in the 1980s. As table 10.1 shows, Shorts had returned to tier 1 activities with a vengeance; offering machines in the 21–40 passenger class at prices bettered only by Spain's CASA C–212. Moreover, it achieved a major breakthrough into military markets in 1984 with an order from the US Air Force for eighteen of a version of its model 330 worth $165 million (sufficient to guarantee 1000 jobs for a year). Meanwhile, the competitiveness of the

Table 10.1 Commuterliner prices, 1984

Company	Product	Price ($ million)
CASA (Spain)	C–212/200	2.4
CSA/PT. Nurtanio		
(Spain–Indonesia)	CN–235	4.9
De Havilland Canada	DHC–8	5.1
Embraer (Brazil)	Brasilia	4.7
Saab–Fairchild		
(Sweden–USA)	SF–340	5.25
Shorts	330	3.5
Shorts	360	4.4

Source: Interavia, May 1984, pp. 424–6

model 360 was vindicated by a £30 million order for eight of the aircraft from China; an order which should lead to follow-on contracts. At about the same time, the company won a fixed-price contract of £125 million to supply 130 Embraer Tucano trainers to the RAF. The fruits of a joint venture between the Brazilian firm and Shorts, the programme should create 1100 airframe jobs in the UK – more than half of which will be in Belfast. Reputedly, this contract went to Shorts rather than other UK contenders owing to the political desirability of maximizing job creation in that province (*Flight International*, 30 March 1985, p. 1). Shorts was equally emphasizing the missile side of its tier 1 portfolio, building man-portable Javelin missiles for the Army and a range of naval surface-to-air missiles for domestic and export markets.

All these efforts required continual injections of government funds over and above the monies found from the company's resources. In 1980, for example, Shorts received $43 million from the UK Government in order to cover losses of $40 million in 1978–9 and finance the development of the model 360. In 1984, the Northern Ireland Department of Economic Development (the company's controlling shareholder and UK Government watchdog) committed £30 million to cover the costs involved in developing the wing for the F–100 airliner: a risk-sharing venture with Fokker which extends the practice of co-operation inaugurated with the F–28 in the 1960s. Clearly, the price of propping up an existing aerospace enterprise is not insignificant and shows little sign of lessening over time despite the formulation of marketable civil aircraft designs by the firm and its ability to garner revenues from selling missiles and trainers to defence customers. In fact, the firm made a profit in 1985 (£530 000) after an unbroken eleven year run of losses (*Flight International*, 30 November 1985, p. 105).

If maintaining an existing firm is costly, the Northern Ireland experience suggests that starting a new one from scratch may be too costly. Centred on the Lear Fan Ltd venture, the new firm was a brave attempt to target tier 1 aerospace for regional development and broaden the existing base provided by Shorts. To be sure, the original US sponsors of the firm had wanted to locate at Bristol and only shifted to Ulster at UK Government insistence. Lear Fan was the brainchild of William Lear (who also designed the original 'bizjet' – the Learjet – and the initial variant of the Canadair Challenger). Lear Fan focused on the model–2100 twin-turboprop business aircraft; a revolutionary design using composite materials and the principle of independent-drive pusher propulsion (DeMais, 1985). Split between US and UK locations, the UK division of the enterprise, Lear Fan Ltd, was a subsidiary of Lear Avia Corporation of Reno, Nevada. The latter was held responsible for R&D and testing, reserving Northern Ireland for the manufacture of production aircraft. Initial capitalization in 1980 was $80 million, of which $50 million derived from UK Government sources ($35 million for industrial grants and loans and the balance for loan guarantees). In 1982, the international venture employed 560 at Carmoney and Aldergrove near Belfast and 400 at Reno. By then, however, it was desperate for development money; a problem which was only overcome by the restructuring of the US parent (as Fan Holdings Inc) with the assistance of Saudi investment. Plans were afoot in the immediate aura of optimism to generate 2800 jobs in Northern Ireland by 1987. For its part, the Department of Economic Development (which retained a five per cent equity stake) promised a further $30 million on the understanding that private investors would be forthcoming with twice that amount. As a result, a 45,000 m² factory at Antrim was procured and steps were taken to produce aircraft components for shipment to Reno as a first phase in the goal of manufacturing the aircraft entirely within Northern Ireland. It was suggested that the UK branch of the operation would represent about 70 per cent of total production man hours.

Unfortunately, structural problems with the airframe of the 2100 encountered in 1984 imposed delays on American certification (a vital requirement if the machine was to be sold in the dominant US market) and led to the lay-off of most of Lear Fan's employees in Northern Ireland. While these difficulties were partially resolved, full certification was never achieved and, accordingly, production plans were abandoned. Attempts at reviving the venture were tried, but proved fruitless as fresh capital failed to materialize. Potential backers were deterred because the technical problems associated with the tooling processes for composite-material airframe manufacture remained to be satisfactorily resolved; added to which were the design difficulties stemming from the craft's unique configuration (Birch, 1984). Lear Fan was plagued, indeed, by the bugbears of a true pioneer: which is to say, it had problems with developing both process and product innovations. It is fair to say, however, that these technical problems were exacerbated by political pressures imposed on the firm to the extent of forcing it to take on 360 Belfast workers well before production scheduling warranted such action. In any event, the difficulties all conspired to close the enterprise in June 1985 with the write off of £57 million in government subsidies.

MANITOBA

The Manitoba Government's 1970 decision to attract a tier 1 manufacturer was stimulated by two factors. On the one hand, there was a real need to boost the province's lagging manufacturing sector. Manufacturing investment in Manitoba had constituted 2.4 per cent of Canada's total in 1966 but was steadily eroding; by 1977 the proportion had dropped to 1.3 per cent (Canada/Manitoba, 1978). On the other, the untimely closing of the Gimli air base left a considerable void in the Interlake regional plan centred on Gimli. Airframe manufacture could, at one fell swoop, make use of the redundant aeronautical facilities and endow the province with a high-tech industry which was not impaired by such an 'eccentric' location. Consequently, Manitoba successfully induced the newly created Saunders enterprise to move from its founding designer/entrepreneur's site at Montreal and occupy the Gimli facilities. The transfer was occasioned by the capital shortage of the firm being made good by the Manitoba Development Corporation (conditional on relocation and an 81.7 per cent equity holding). Initially, Saunders produced the ST–27, a turboprop remanufacture of the UK Hawker-Siddeley Heron 23-seat commuterliner. By mid-1975, employment had built up to 450 and twelve ST–27s had been completed.

In view of uncertain sales prospects, the Manitoba and Saskatchewan governments committed themselves to the purchase of two ST–27s for a local interprovincial commuter service. But while the machine received UK certification, it ran into difficulties in obtaining the all-important US certification and therefore was denied access to the US market. To offset these difficulties, Saunders pursued a sales prospect in Chile which, if successful, may have secured the aircraft's future. However, the firm found itself caught between conflicting federal and provincial industrial strategies. Competing against Saunders for the Chilean sale was the federal government's newly acquired de Havilland Canada company (DHC) with its Twin Otter aircraft. Whereas the Chile sale may have been enough to secure some 400 jobs at Saunders, DHC was an established, albeit ailing, corporation whose Twin Otter had realized 340 sales by January 1972. Given the federal government's

interest in DHC and its location in Toronto, political leverage clearly favoured
the central Canadian firm. At any rate, the federal government's Export
Development Corporation preferred to back the sale of the Twin Otter. In
the absence of a 'soft' financing package, as was extended to DHC (no
payments for five years and an additional five years to pay back the loan),
Saunders was unable to compete and thus lost the sale.

The ST–27 was intended as the primary stage of Manitoba's aircraft
development programme. The second stage, the ST–28, was an incremental
innovation geared to the same commuter market. Yet, as with its predecessor,
the ST–28 encountered US certification problems and the company estimated
that C$10 million was required simply to overcome them. Disillusioned by
this scenario and the fact that C$37 million had already been expended,
the Manitoba government closed down the Gimli operation in December
1975. The federal government was unconcerned by this outcome. It had
pledged some C$8 million for Saunders (to come in equal amounts from the
Department of Regional Economic Expansion and the Program for the
Advancement of Industrial Technology), but this was contingent upon the
successful certification of the ST–28. In making support conditional on
certification, Ottawa effectively sidestepped responsibility for the programme's demise. Such indifference to regional aerospace was articulated
by a federal task force which stated that governments 'should not attempt
a dispersal of the core of the industry from the present centres (in central
Canada) if this requires subsidization to obviate the introduction of
uneconomic factors of marketing and production' (Canada, 1978, p. 11).
Instead, the regions should be content with tier 3 activity.

Given the failure of Saunders and in view of the declared position of
Ottawa, the province subsequently concentrated on attracting tier 2 and 3
activities. Three were successfully induced to locate in Winnipeg, effectively
doubling the province's complement of aerospace enterprises. Of the total
of six firms, four are wholly owned foreign subsidiaries, one is central
Canadian controlled, leaving a single local enterprise. These developments
owe much to regional policy jointly undertaken by both federal and provincial
levels of government. The main instrument used in this respect has been
defence procurement. Thus, two of the induced firms are the products of
industrial offset arrangements negotiated between Ottawa and US prime
contractors as part of major Canadian purchases of US combat aircraft (that
is, a Boeing composite materials plant and a Sperry-Univac avionics factory).
In other words, the plants were located in Canada in return for Canadian
orders for American defence equipment. Paradoxically, while both these
plants were conceived as industrial offset benefits to Canadian industry, they
were subsidized to locate in Manitoba at public expense. In fact, all six of

Manitoba's aerospace firms have received grants to the tune of C$5.4 million from the Regional Development Incentives Act scheme. For that expense, some 752 direct jobs were created at a cost of C$7,200 per job.

All told, some C$8.3 million of industrial assistance has been extended to Manitoba's current aerospace firms. In recompense, the industry remains a relatively important part of the province's industrial fabric, employing more than 2500 and generating sales in excess of C$250 million. The benefits are expected to increase until, in 1987, the industry will directly employ 3000; it will indirectly add a further 2500, and will realize sales in the order of C$320 million (table 10.2). Moreover, the industry is highly export oriented and will become even more so in succeeding years. In so far as a sizeable proportion of these exports are destined for the US market and are thus

Table 10.2   Manitoba aerospace benefits (estimates)

Year	Direct employment	Total sales ($ m)	Exports ($ m)	Indirect employment	Total wages of direct and indirect employment ($ m)
1984	2506	223	124	1877	94
1985	2601	249	140	1956	104
1986	2791	281	162	2107	117
1987	3006	327	199	2261	138

Source: Economic Research & Analysis Branch, Manitoba Dept. of Industry, Trade and Technology

covered under the Canada/US Defence Production Sharing Agreement, the export vitality of the industry is highly sensitive to levels of defence expenditure: a seeming truism affecting all aerospace industries be they tier 1, 2 or 3, targeted or otherwise.

THE REGIONAL AEROSPACE OPTION

The cases of Northern Ireland and Manitoba represent in microcosm the advantages and disadvantages of promoting regional aerospace ventures. The experiences of other regions, summarized in table 10.3, mirror those highlighted in the two subjected to enquiry. The first point of note is the functional option. Thus, in contradistinction to Northern Ireland, the Manitoba authorities have latterly fostered tier 2 and 3 functions. For their efforts, they have obtained 2500 jobs (excluding indirect/induced jobs) for a relatively small outlay, whereas the 6000 or so tier 1 jobs maintained in Ulster are bought at a much higher (albeit hidden) public subsidy. Opting for tier 2 and 3 specialization does, however, carry a cost; namely, an abrogated R&D commitment. In other words, circumventing the high-risks/high-cost equation of tier 1 means that much of the technology-intensive part of aerospace is omitted as well, as are the corresponding spin-offs into other industries. Of course, many tier 2 enterprises are research intensive and suffice to offer a region such as Manitoba at least the foundations of high-tech industry at reduced levels of both cost and risk. However, the fundamental dilemma remains: R&D commitment is generally restricted as a direct result of the subsidiary status adhering to tier 2 activities. They are, in effect, elements of the disparaged 'branch-plant syndrome'. The only seeming alternative for regions – public or quasi-public enterprises – may be unacceptable on the grounds of cost (e.g. Saunders) or politics (e.g. the proposed privatization of Shorts).

Ownership notwithstanding, the second notable feature of regional aerospace is its heavy reliance on technology transferred from elsewhere. As well as the Lear Fan and Saunders cases described, a motley range of regions have attempted to induce 'outsider' designer/entrepreneurs to transfer to their jurisdictions (table 10.3). The results have been generally unimpressive. In Italy, excluding the Mezzogiorno which receives a quota of the investment made by nationally operating state enterprises, the other regions have relied on packages of inducements offered by regional authorities. In spite of such fiscal and marketing assistance, virtually all regional ventures have failed within a short time of foundation, more often than not before a viable product has materialized. Like Lear Fan, the Belgian Foxjet was an American-inspired revolutionary new 'bizjet' which failed to reach certification. Like Saunders, similar ventures in Ohio and Puerto Rico hinged on low-cost commuterliner projects whose promise, as it happened, never reached fruition. In an elusive search for civil markets, commuterliners are particularly apt for targeting (possessing relatively low technical barriers to entry). Their prototype in a targeting sense was the Potez scheme in the Irish Republic. Conceived in the early 1960, the French Potez firm established a production line at Baldonnel in order to build the model–840 commuterliner which was expected to do well in the US market. Failure to live up to expectations led, in 1966, to the write off of the project after Potez had put up £3.5 million and the Irish government had subscribed an additional £1 million. Despite claims of employment amounting to 1500, the actual workforce never exceeded 70. Undaunted by such inherent risks, some regions retain faith in the targeting of tier 1 activities which are transferred from elsewhere. Three recent examples endorse that assertion. In the USA, the state of Missouri advanced $5 million for the erection of a plant at Harrisonville in order to host production of the Skytrader commuterliner. Before locating in Missouri

Table 10.3 Regional aerospace ventures

Location	Enterprise	Backer	Form of support	Outcome
Mezzogiorno	Aeritalia Agusta (state firms)	Italian Government	Commitment to invest in region	Plant extensions in Naples area Brindisi airframe plant Frosinone helicopter plant
Flanders	Flemish Aerospace Group	Belgian Government and private interests	Commitment to purchase defence items from region	Plant to manufacture US-designed Foxjet business aircraft at Zutendael (abandoned 1985?) Promised site for helicopter factory
British Columbia (Canada)	Trident Aircraft	BC and Canadian Governments, private interests	launch costs plus factory costs	Enterprise failed in 1980 through lack of support, Sidney factory incomplete
Québec (Canada)	Avions Robin Canada	48% equity from Quebec Govt., 25% plant costs from Canada	Facilitate establishment of subsidiary of French light aircraft firm	Plant assembling French-made aircraft at Lachute
Ohio (USA)	Commuter Aircraft Corp.	Ohio State Govt. and US Dept. of Commerce	loan guarantees	Firm failed, Youngstown airframe plant incomplete
Puerto Rico	Ahrens Aircraft	PR Government	tax concessions, arranged financing	Firm failed, original entrepreneurs attempted to relocate Sweden

in 1984, the Skytrader had tried locales in Washington, British Columbia and Alberta without success. Also in the US, the state of New Mexico found $20 million to support development of the Avtek–400 business aircraft. The State was captivated by the prospect of 750 jobs by 1987 and a 'Kevlar Valley' of composite-material manufacturers on the outskirts of Albuquerque. It was ready to overlook any technical and marketing risks associated with this California design. In a slightly different vein, the Welsh Development Agency persuaded the NDN firm to shift from the Isle of Wight to Barry (and eventually Cardiff Airport) in return for generous backing. Focusing on trainer and agricultural aircraft, this venture (restyled NAC) attempted to emulate NICs such as Brazil in entering the less-sophisticated end of the spectrum of tier 1 activities. Whether it will succeed is a matter of speculation, but the contemporaneous failure of an equivalent innovative firm (Edgley) and the inability of NDN to attract RAF interest in its trainers, are not good auguries. In promoting the designs of NDN's founder, Des Norman, the Welsh venture is at least attempting to foment indigenous and original R&D. The same cannot be said for the attempt in Quebec to develop light aircraft manufacture. The Quebec operation is little more than an assembly operation of French and US parts and it conforms to the same abrogated R&D pattern as was highlighted in the case of Manitoba's tier 2 strategy.

A third notable feature of aerospace targeting is the lumpy capital requirement. Massive capital injections are normal in aerospace and hinge, in the final analysis, on the willingness of governments to provide the wherewithal for advancing technology. A coherent defence procurement policy is helpful in this respect. A pitiful civil government purchase of ST–27s effectively doomed Saunders, while Lear Fan could not rely on defence funding to cross-subsidize its heavy, and ultimately insurmountable, development costs. Conversely, defence orders sustain Manitoba's tier 2 plants and UK Ministry of Defence (with some help from the USAF) contracts ensure the future of Shorts. Similarly, aerospace activities in Italy's south are, by and large, underwritten by military work. Quite simply, regional aerospace industries are unlikely to prosper without the kind of guarantees afforded by defence ministries. Demand management through public procurement reduces, in the main, to judicious allocation of defence contracts among regional aerospace firms (Lovering 1985). As with other high-tech sectors, appropriately construed public procurement could work wonders in allowing neophyte regional aerospace firms to overcome teething troubles (Jeanrenaud 1984, Rothwell 1984).

CONCLUSIONS

The hallmark of aerospace as a target industry is its dependence on technology transfer. Irrespective of whether it is subsumed under such expressions as design, entrepreneurship or product or process innovation, technology-transfer processes are essential. It is hardly remarkable, therefore, that attempts to promote regional aerospace have been unable to shrug off the dependency aspect. In Northern Ireland, for example, Shorts relies on US engines for its commuterliners while Lear Fan was conceived from the beginning as merely the production arm for US technology. Manitoba has apparently resigned all ambitions for nurturing an indigenous R&D posture in aerospace. Similarly, by virtue of its willingness to undertake licence production of Brazilian aircraft, Ulster has tacitly come to accept its aerospace sector as a job provider rather than as a crucible for high-tech spin-offs into the local economy. Undoubtedly, any pretensions to creating technologically vibrant aerospace enterprises carry with them a high-risk tag which translates into sizeable, and apparently permanent, state subsidies. Ironically, avoidance of the high-risk/high-cost equation means abandonment of the fruits of meaningful technological spin-offs and reversion to *de facto* branch-plant status. In short, targeting aerospace for the regions does risk reinforcing the fragile, unstable nature of their economies, a prospect scarcely in tune with the heroic claims of its supposed high-tech potential. Indeed, only by treating regional aerospace as a 'national' instrument for support will the resources be forthcoming to ensure its technological viability. Severance of the economic dependency status of regions requires nothing less than the total integration of aspatial and regional industrial policy and, in the case of aerospace targeting, of defence policy as well.

REFERENCES

Adams, F. G. and Klein, L. R. (Eds.) (1983) *Industrial Policies for Growth and Competitiveness: An Economic Perspective* Lexington Books, Lexington, Mass.
AIAA (1985) *Aerospace Facts and Figures*, Washington, DC.
Birch, S. (1984) Composite problems at Lear Fan, *The Engineer* 7 June 14–15.
Canada (1978) *A Report by the Sector Task Force on the Canadian Aerospace Industry*, Ottawa.
Canada/Manitoba (1978) *Subsidiary Agreement: Industrial Development*, Ottawa.
Clarke, I. M. (1982) The changing international division of labour within ICI. In Taylor M. and Thrift N. (Eds.), *The Geography of Multinationals*, Croom Helm, London, 90–116.

Corlett, J. (1981) *Aviation in Ulster*, Belfast.

DeMais, W. (1985) Agonies of the Lear Fan, *Aerospace America* 24 (October), 52–60.

Froebel, F., Heinrichs, J. and Kreye, O. (1980) *The New International Division of Labour: Structural Unemployment in Industrialised Countries and Industrialisation in Developing Countries*, Cambridge University Press, Cambridge.

Jeanrenaud, C. (1984) Public procurement and economic policy, *Annals of Public and Co-operative Economy* 55, 151–158.

Klein, B. H. (1977) *Dynamic Economics*, Harvard University Press, Cambridge, Mass.

Lipietz, A. (1983) Towards global Fordism?, *New Left Review* 132, 33–47.

Lovering, J. (1985) Regional intervention, defence industries and the structuring of space in Britain, *Environment and Planning D* 3, 85–107.

Magaziner, I. C. and Reich, R. B. (1982) *Minding America's Business: The Decline and Rise of the American Economy*, Harcourt Brace Jovanovich, New York.

Malecki, E. J. (1983) Technology and regional development: a survey, *International Regional Science Review* 8, 89–125.

MITI, (1982) *Japanese Aerospace Prospects*, Tokyo.

OECD, (1979) *The Impact of the Newly Industrialising Countries on Production and Trade in Manufacturing*, OECD, Paris.

Pavitt, K. (Ed.) (1980) *Technical Innovation and British Economic Performance*, Macmillan, London.

Rothwell, R. (1984) Creating a regional innovation-oriented infrastructure: the role of public procurement, *Annals of Public and Co-operative Economy* 55, 159–172.

Thurow, L. C. (1984) The need for industrial policies: the case of the USA, *Annals of Public and Co-operative Economy* 55, 3–31.

Todd, D., Simpson, J. and Humble, R. (1985) *Aerospace and Development: A Survey*, Winnipeg.

11

Innovation policy and mature industries: the forest product sector in British Columbia

R. Hayter

Since 1970, science and technology policies for industry have emerged as an important, possibly dominant, theme in the national and regional economic policies of virtually all industrialized countries of the western world (Rothwell and Zegweld, 1981; Schott, 1981; Vos, 1983). Heightened awareness of the importance of technology for economic growth, increasingly severe recessionary conditions, the rapid emergence of newly industrialized countries and of Japan as the world's leading industrial power, and the energy crisis of the early 1970s, as well as frustration with conventional industrial stimulation programmes, have encouraged governments to elevate innovation policy within their overall economic thinking (Ewers and Wettmann, 1980). Indeed, most governments in developed countries consider their economies to be undergoing profound secular change and recognize that innovation is essential for the maintenance of industrial competitiveness and the creation of employment opportunities (Freeman, Clark and Soete, 1982).

Despite considerable variations in content, commitment and perspective, the central thrust of innovation policies (and discussions) has been the development and attraction of newly emerging businesses in so-called 'high technology industries'. Admittedly, the distinction between 'high tech' and 'low (and medium) tech' industries is not always clear. There is a reasonably broad consensus, however, that 'high tech' industries primarily, if by no means exclusively, comprise activities which are research intensive as measured by the relative importance of scientists and engineers to overall manpower requirements or by research and development (R & D) budgets as a percentage of sales. In practice, the microelectronic industry is regarded as the quintessential 'high tech' activity of the 1970s and 1980s which is providing the basis for the rapidly evolving information age and driving economic growth across a broad front by spawning both significant productivity advances and product innovations throughout the manufacturing and service sectors.

Certainly, microelectronics is at the core of contemporary science and technology policies for industry (Nelson, 1984; Vos, 1983). At the national level, in fact, the quite different policies of the US and Japan towards the stimulation of 'high tech' activities, notably microelectronics, have established their technological leadership and have set the standard for the more recent innovation policies of other advanced countries. It is interesting to note that even governments which have been traditionally 'non-interventionist' in their approach to economic policy have enthusiastically adopted new policies and programmes to stimulate innovation in the high technology sector (Freeman, 1978, p. 6). At the regional level, the proliferation of innovation policies throughout Europe (Ewers and Wettman, 1980) and North America (Joint Economic Committee, 1982) have also replicated the national preference for 'high tech' activities, notably microelectronics. In this context a widespread policy tool has been the creation of innovation centres or parks which have been primarily conceived of as imitations of Silicon Valley and the Boston 128 Complex (see chapter 12). The Research Triangle Park of North Carolina and Ottawa's technology-oriented complex, both of which are oriented to microelectronics, are two of the more successful initiatives in this regard (Steed and DeGenova, 1983; Whittington, 1985).

In contrast, mature (or 'low and medium tech') industries such as textiles, steel, shipbuilding, lumber and even automobiles in some instances, hitherto have rarely been explicitly incorporated within the framework of innovation policy. The long standing recommendations of the Science Council of Canada that the resource industries be established as innovation priorities are exceptional in this regard (Britton and Gilmour, 1978); in any event, these pleas have not found practical expression. In Canada, as in several other advanced countries, including the UK, national and regional policy debates about mature industries typically concern whether to halt, slow down or accelerate the (actual or perceived) rate of decline. Moreover, discussions of the pros and cons of government support for mature industries, whether indirectly in the form of (say) tariff protection or directly in the form of grants, have increasingly emphasized social rather than economic arguments. As such, they have become fundamentally defensive in nature.

At least within Anglo-America, it is felt that the distinction between 'high tech' and 'low tech' industries, and between the 'old' and the 'new' economy (Cohen and Shannon, 1984), is an increasingly important feature of national and regional economic planning. Indeed, within the related literature there is a similar tendency for studies of research, development and innovation to stress youthful industries. Perhaps even more importantly, the 'low tech/high tech' dichotomy has been encouraged by contemporary long wave, life cycle and innovation cycle theories. Less recognition is given to the fact, however, that these theories are strongly idealized and, in many respects, unsubstantiated.

From an innovation policy perspective, there are legitimate reasons encouraging governments to focus on 'high tech' activities (Ewers and Wettman, 1980). Whether such policies will provide a 'quick fix' to economic problems is more debatable. 'High tech' oriented innovation policies, for example, face considerable uncertainty. In this regard, Nelson's (1984) comment that the record of countries attempting to play technological catch-up to the US and Japan has been one of 'expensive frustration' is probably a fair reflection of regional experience. In addition, there are serious questions about how many new jobs a narrow range of 'high tech' activities might be expected to generate over the next decade or so. On the other hand, there is increasing evidence that innovation is the key to the rejuvenation of many mature industries (Abernathy, et al., 1983; Piore and Sabel, 1985; Rees, Briggs and Hicks, 1985). That is, mature industries do have technological options and whether or not they become part of a sunset sector is frequently a matter of choice. Moreover, explicit technology policies for mature industries could normally be directed towards peripheral regions which many observers argue are disadvantaged by prevailing high tech innovation policies.

It is the purpose of this chapter to question the wisdom of regional innovation policies which are preoccupied with 'high tech' activities to the exclusion of mature industries. The argument is explored with specific reference to the Province of British Columbia which provides a good example of how, even in a peripheral resource-based region, policy makers have essentially equated innovation policy with 'high tech' activities, while policies for the established industrial base, in this case principally the forest industries, are primarily thought of as, at best, maintaining the status quo. This chapter does not necessarily question the ambition of the Province for diversification into technologically sophisticated activities. It is argued, however, that there are sound reasons for establishing the forest industries as an innovation policy priority. The idea of the forest industries (and other mature industries) comprising a sunset sector following an inevitable path of decline needs to be critically reviewed.

MATURE INDUSTRIES AND INNOVATION-CYCLE THEORY

Reference, or at least widespread reference, to the idea of youthful and mature or sunrise and sunset industries is relatively recent. Although these terms are rarely defined explicitly, the distinction between youthful (sunrise) and mature (sunset) industries rests on a life-cycle interpretation of industrial (and sectoral) evolution. In practice, the classification of an industry's development into such life-cycle stages as birth, youth, maturity, old age and death, is ultimately based on technological considerations. Indeed, the distinction between youthful and mature industries is often popularly, if loosely, associated with so-called 'low technology' and 'high technology' activities. More precisely, the idea of maturity within the context of industrial development has been defined as a 'process by which competition becomes progressively immune to technology-based change from within the industry' (Abernathy et al., 1983, p. 27).

A widely cited technology-based model of industry evolution has been developed by Utterbach and Abernathy (1975; see also Abernathy and Utterbach, 1978). According to Utterbach and Abernathy (1975), as industries mature the underlying nature of major innovations changes from a focus on new product development in early 'fluid' stages of development to one of process optimization and cost reduction in later 'specific' stages of development. In the fluid pattern of development this model suggests organizational control is informal and entrepreneurial, plants are small scale, production processes are flexible and the products are diverse and frequently changed. During the 'specific' pattern of development, the competitive emphasis is on cost reduction rather than product performance, plants are highly efficient, capital intensive, high cost, large, and specialized and they produce undifferentiated standard products. It is argued that, parallel with these shifts, the nature of investment changes from a 'net expansionary mode into a net rationalization mode' (Rothwell and Zegweld, 1981, p. 42).

This innovation theory of industry evolution has strong affinities to the product-cycle model developed by Vernon (1966) and others (e.g. Hirsch, 1967; Wells, 1968) to explain America's trade in consumer durables (Abernathy, et al., 1983). Innovation-cycle theory also relates to the more macro 'long wave' theories of economic growth reviewed by Freeman in chapter 2. These suggest that the industrial crisis of the 1980s is the last of a series of increasingly severe recessions which characterize the downward swing of forty or fifty year long (Kondratiev) cycles. According to Mensch (1979), for example, in a development of the Schumpeterian view of long wave theory, the situation facing the industrialized countries since the latter part of the 1970s has been one of 'technological stalemate'. Briefly stated, it is argued that the massive opportunities for investment and employment in new branches of industry which were generated by a cluster of radical innovations in the 1930s have now played themselves out. The main thrust of remaining investment is to increase capital intensity and to economize on labour and materials. Ultimately, escape from stalemate depends upon the occurrence of another cluster of radical innovations which are employment generating.

Within the advanced countries, mature industries may therefore be thought of as those that were established one or more Kondratiev cycles ago and which utilize known technologies to manufacture standard commodities of low value per item or per unit weight (or sometimes more ambiguously as activities which do not add much value). In this context, sunset industries can be best regarded as those mature industries whose decline, as measured by employment, output levels, sales etc., is underway or imminent because of, for example, reduced levels of demand, changing consumer tastes, the development of substitute products or increasingly inappropriate supply or location conditions such as wage levels, taxation levels and raw material availability.

There is no question that contemporary economies are experiencing deep-seated structural transformation and that traditional ('mature') industries will not provide the *same* kind of platform for growth as they did in the past. Indeed, in some cases, they will decline absolutely. In this respect, innovation-cycle theory, and other life-cycle theories, is extremely useful for stressing the secular nature of economic change. Yet it is important to bear in mind that innovation-cycle theory, as conventionally represented, is strongly idealized and presents a deterministic view of technology-based secular change.

The strongest criticism of innovation-cycle theory is that it underestimates the technological dynamism of mature industries. The theory does implicitly recognize that major innovations spawn a whole series of incremental innovations which, in the case of process technology, can generate productivity impacts cumulatively greater than the initial major innovations (Hollander, 1965). An inherent characteristic of technological change is its uncertainty, however, and there is no reason to accept automatically the assertion that within an industry the rate of *major* innovations declines smoothly or that the relationship between product and process innovation should inevitably change in a constant way over time. Indeed, if it is reasonable to suppose that basic innovations 'cluster' at periodic (and unpredictable) intervals in the economy as a whole (see Freeman et al., 1983), then similar tendencies can be postulated for individual sectors and industries. Chapters seven and eight by Holmes and Bradbury are consistent with a mounting body of opinion and evidence that rejects the inevitability of the maturity thesis.

Even the original sponsors of innovation cycle theory now wish to reject the idea of the maturity thesis as an irreversible process. Thus, Abernathy et al., remain sympathetic with an evolutionary view of technological change but 'with its dogmatic acceptance as an inflexible and unchangeable model of the way things are and must always be, we have our doubts' (Abernathy, et al., 1983, p. 19). Indeed, in the case of the American automobile industry, these authors identify four peaks of major innovation activity which began in the 1890s, the early 1920s, the late 1950s and the mid 1970s, each of which had significant 'transilience effects', that is, impacts on established systems of production (Abernathy et al., 1983, pp. 114–18). They also point to the possibility of major technological advances in the future both in terms of improvements to existing designs and the innovation of such 'revolutionary' concepts as (disposable) plastic engines, electrically powered cars and hydrogen-powered turbo-engines. They suggest that innovations along these lines are critical to the prosperity of the American automobile industry.

Similarly, but on the basis of a broader range of historical and international evidence, Piore and Sabel (1985) argue against thinking in terms of 'natural technological trajectories'. Using examples from the steel, textiles and cutlery industries, among other mature industries, they contrast the technological options that were historically chosen by different regions and emphasize the technological options available at the present time. In particular, extension of the mass production model, the predominant technological option exercised by advanced countries over the past 200 years, is compared with technological responses that reflect 'flexible specialization'. Thus flexible specialization is defined (and exemplified) as

> a strategy of permanent innovation: accommodation to ceaseless change, rather than an effort to control it. This strategy is based on flexible-multi-use-equipment; skilled workers; and the creation, through politics, of an industrial community that restricts the focus of competition to those favouring innovation. (Piore and Sabel, 1985, p. 17)

For Piore and Sabel, possibilities for the rejuvenation of mature industries fundamentally depend upon various strategies of flexible specialization. In this regard, a critical question clearly concerns the ability of mature industries characterized by long established technological conservatism to develop the institutions and attitudes necessary to pursue more innovative strategies. One approach to the breakdown of such conservative attitudes, identified by McArthur in chapter 3, is the reorganization of production into smaller units – a trend which has been facilitated in certain sectors by the development of computer-controlled machinery.

It should also be noted that innovation-cycle theory, in conceptualization, ignores inter-industry and inter-sectoral linkages and, in so doing, implies that youthful (sunrise) and mature (sunset) industries are independent of one another. Yet, there is a significant literature which stresses the role of industrial linkage mechanisms in long run processes of regional industrial change. In the case of export base theory, for example, regional growth is envisaged as a diversification process which occurs largely as a result of the generation of forward, backward and final demand linkages around the initial export (or staple) base (Watkins, 1963). From this perspective, the industries of the future are likely to be strongly shaped by past processes of industrialization, so that policies to stimulate newly emerging 'high tech' businesses clearly ought to be at least cognizant of the technological capabilities and needs of the existing industrial base.

The whole question of how priorities are identified within the context of innovation policy is an extremely problematical one (Mensch, 1979). There are reasons to believe, however, that governments, especially in peripheral regions, should give greater thought to integrating mature industries within their innovation policies and as a matter of priority. The possibilities in this regard will clearly be closely contingent on particular circumstances. Accordingly, the remainder of this chapter considers the case of the forest (product) industries as an innovation priority in British Columbia. As an introduction it might be noted that the forest industries have been the Province's most significant source of employment and income throughout the twentieth century. Lumber and market pulp, in particular, are world-scale industries and, in association with logging, plywood, other wood-processing activities, and newsprint and converted paper manufactures, in 1974 they were estimated as directly employing about 100 000 persons while accounting for 50 cents of every dollar generated in the Province (Reed, 1975). Nevertheless, government planners have increasingly perceived forest products as a sunset sector rather than as a central focus for innovation policy.

FOREST PRODUCTS AS A SUNSET SECTOR IN BRITISH COLUMBIA

Within British Columbia the idea of the forest industries as a sunset sector emerged gradually following the energy crisis of the early 1970s when it was recognized that the industries' growth prospects had changed. The energy crisis itself did not disadvantage the forest industries in British Columbia as much as elsewhere. Yet it did herald a period of deteriorating supply and demand conditions and while record production (and employment) levels in all major forest product commodities were recorded in 1979, rates of growth of these commodities during the 1970s were substantially lower than in the 1950–70 period (Hayter, 1976). That is, the post-war boom 'levelled-off' and the era of dynamic forest product growth based on new mills exploiting new timber supply areas was over.

There were other signs of disquiet during the 1970s. Technological change elsewhere increasingly eroded British Columbia's natural high quality resource advantage, the strength of competition from the southern US became more apparent, major customers increased their levels of self sufficiency, and concern for the adequacy of the resource base began to be voiced more strongly. Nevertheless, the boom years of the late 1970s encouraged forest product firms to embark upon very expensive expansion and modernization schemes. Unfortunately, these capital investments, often financed at high and increasing rates of interest, were not even completed when the worst recession in fifty years began in 1980 and serious questions about the viability of the forest industries were publicly voiced.

The impacts of recession

The effects of the recent recession on the forest product industries of British Columbia have certainly been harsh. In 1981, for example, the forest industries incurred a massive $500m loss and generated a return on capital of just 0.6 per cent; in 1979 the industries had generated a profit of $500m and a return on capital of 8.9% (Farris, 1983, p. 49). Every important company which publicly reported financial performance incurred losses in the early 1980s and in several instances losses were recorded in three consecutive years. In addition, debt-equity ratios reached unusually high levels and, in key commodities, production levels in 1981 and 1982 dropped significantly below the 1979 record levels.

As would be expected under these circumstances, job losses were substantial and rapid. One conservative estimate, for example, based on federal government sources, suggests that between 1979 and 1982 job loss amounted to almost 23 000 (table 11.1). Despite the fact that production levels recovered substantially in 1983 and 1984, prices remained depressed and there is little doubt job losses continued. Some insights into the nature of job losses is provided by a cursory examination of the employment changes implemented by two of the largest forest product corporations in their manufacturing facilities between December, 1980 and December, 1984 (table 11.2).

For both these firms, modernization and rationalization have provided important production contexts for job loss. At Chemainus, for example, MacMillan Bloedel replaced an old large-log sawmill employing 682 people in 1980 with a highly automated facility of almost the same capacity but which employs only 100 people. Similarly, at Fraser Mills, Crown Forest (formerly Crown Zellerbach) reduced its employment in its lumber and plywood

Table 11.1 Employment change in the forest product industries of British Columbia 1979–1982

Industry	1979	1980	1982	Job change 1979–1982
Logging	24 474	24 270	18 000	− 6 474
Lumber	37 257	35 850	29 000	− 8 757
Shingles	1 680	1 714	1 300	− 380
Plywood & veneer	7 930	6 928	5 400	− 2 540
Other wood	4 502	5 216	3 800	− 702
Pulp and paper	18 600	19 100	15 700	− 2 900
Misc. paper	3 400	- 2 900	2 300	− 1 100
				− 22 853

Source: Farris, 1983 (and based on Statistics Canada: Canadian Forestry Statistics). Note that activities related to reforestation, forest fire fighting and nurseries are excluded

Table 11.2 Employment changes in selected manufacturing facilities in British Columbia of two large forest product firms

Location	Product	Employment 1980	1984	Net Change 1979–1984	Context
MacMillan Bloedel					
Vancouver	head-office	1200	560	− 737	reorganization[a]
Chemainus	lumber	682	100	− 562	modernization
Vancouver	lumber	740	400	− 340	rationalization
Vancouver	plywood	400	0	− 400	rationalization
Vancouver	particleboard	91	70	− 21	intensification
Port Alberni	plywood	450	375	− 75	—
Port Alberni	lumber (two mills)	1600	950	− 650	modernization[a]
Port Alberni	pulp, paper	1522	1249	− 273	modernization
Powell River	pulp, paper, lumber	2335	1598	− 737	modernization
Harmac	pulp/lumber	1399	1347	− 52	intensification[a]
New Westminster	corrugated containers	161	156	− 5	—
Burnaby	paper bags	94	94	0	—
New Westminster	fine paper	222	372	+ 150	expansion
Crown Forest (formerly Crown Zellerbach Canada)					
Vancouver	head-office	260	190	− 70	reorganization[a]
Richmond	lumber	145	0	− 145	rationalization
Armstrong	lumber	400	360	− 40	intensification[a]
Lumby	lumber	111	99	− 12	intensification[a]
Coquitlam	lumber/plywood	1100	600	− 500	modernization & rationalization
Elk Falls	pulp, paper, lumber	1320	1291	− 29	intensification
Kelowna	paper boxes	79	79	0	—

Source: Author survey, 1985
[a] Job changes also affected by cutbacks caused by poor markets and decreased production

operations from 1100 to 600 as a result of a sawmill modernization programme and the elimination of certain functions (e.g timber drying). At the Powell River pulp and paper mill, one of the oldest in the Province, a considerable amount of labour shedding also occurred, mainly as a result of modernization; otherwise pulp and paper mills have shed labour through intensification and minor modernization schemes. The two firms have also rationalized their operations through plant closure, especially in wood processing. Finally, it might be noted that both firms were acquired by corporations based outside British Columbia and during subsequent reorganization plans head-office employment levels were sharply reduced. Clearly, this information suggests a significant element of the employment decline in the forest industries of British Columbia is of a permanent nature.

It may be argued that the problems of overcapacity, growing concern over cost conditions, significant price reductions and job reducing investments were heightened by the recession and were to be expected in an industry in the late maturity stage. Moreover, during the peak of the recession, informed observers expressed fears that imminent timber shortages would cause further capacity cutbacks and higher raw material costs (e.g. Marchak, 1983). In addition, protectionist sentiments within the US were directed towards the British Columbia lumber industry on the basis that the latter's exports are subsidized and causing irreparable damage to American firms. In fact, the possibility of US quotas on British Columbia lumber imports remains a very real threat. Given the preoccupation of forest product firms simply with survival it is not surprising that the view of the forest product industries as a sunset sector should gain credence.

Response of the provincial government

The increasingly pessimistic view of the forest product industries' future, made explicit by the recession, is seemingly shared by the Provincial Government of British Columbia. On the one hand, its policies towards the forest product industries have been defensive in nature. On the other hand, in the formulation of its innovation policy, which is seen as creating the economic base of the future, the forest industries (and resource industries in general) are neglected.

Although there were prior initiatives, notably the creation of the British Columbia Research Council in 1944, the provincial government's innovation policy for industry is recent. It began formally with the establishment of the Science Council of British Columbia and, more particularly, with the creation of Discovery Foundation and Discovery Parks Incorporated in late 1978. Thus the function of the Science Council is to advise the government on matters dealing with science and technology, and Discovery Parks seek 'to encourage the establishment and growth of scientific, technological and industrial research skills in British Columbia with application to both local and world markets, in an attempt to broaden and strengthen the provinces economic base' (Hayter and Gunton, 1984, p. 29).

At present, there are four Discovery Parks located on various campuses within the Vancouver–Victoria metropolitan area. As the Ministry of Science and Technology stated them, the immediate objectives of Discovery Parks are to stimulate development of high-technology industries within the province; encourage the growth of R & D; promote interaction between tenants of research parks and the students and faculty in adjacent learning institutions; provide employment for graduate students; and to provide a bridge between scientific innovation and industrial application of high technology. In a further development of its 'high tech' aspirations, the government in its 1985 budget statement expressed plans, which depend upon cooperation from the Federal Government, to create enterprise (industrial free trade) zones. While the discovery parks and (planned) enterprise zone initiatives appear to have been conceived independently from one another, both policies focus on the Vancouver–Victoria metropolitan region and both attempt to promote internationally competitive 'high tech' activities.

There is no doubt that the underlying rationale for the provincial government's innovation policy is a determination to establish 'high tech' specialisms in order to stabilize employment and reduce dependence on the resource industries at a time when the latter are thought to be at least in relative decline (Gaudry, 1976). That is, the resource industries were quite deliberately excluded from provincial government thinking about innovation policy. Even existing research organizations which serve the resource industries, such as the British Columbia Research Council, were not perceived as part of the new policy thrust. Rather the government's policy initiatives towards the resource industries over the last few years have been entirely defensive. For example, the government introduced a Commissioner of Critical Industries (which were stated to be the resource industries) apparently in order to help ailing firms develop arguments to justify financial baleouts. So far, the Commissioner's activities have been directed towards forest product firms. In addition, and specifically on behalf of the sawmill industry, the provincial government has been actively supporting the industry in lobbying the US government *not* to introduce tariffs against the export of construction grade lumber from Canada. The provincial government has also negotiated a forestry agreement with the federal government finally (and hopefully) to fund an effective sustained yield policy (and only in the face of dire warning, about imminent reductions in forest harvest levels). Although distinct policies for existing and new businesses are to be expected, the government's initiatives towards the resources industries, including the forest product industries, are essentially defensive, status quo policies. A strong case can be made, however, for the forest product industries being an innovation policy priority.

FOREST PRODUCTS AS AN INNOVATION POLICY PRIORITY

According to conventional innovation-cycle theory technological change plays a decreasingly important role in the competitive process as industries mature. In contrast to this view, there is a growing appreciation, not necessarily shared by the industry's decision makers, that the problems of the forest industries of British Columbia stem from too little (and misplaced) technology rather than from too much, and that an increased commitment to technological innovation and excellence is essential if these industries are to maintain and enhance their contribution to the provincial economy. The basis of this argument is that the forest product industries of British Columbia, to an even greater degree than elsewhere in Canada, have historically developed as a *marginal* supplier of a few standardized bulk commodities using *established* technology and passively responding to the needs of a few metropolitan powers. In fact, the province's forest product industries, despite their size, are dominated by the production of just two commodities: construction grade lumber, overwhelmingly for the US market, and kraft pulp, mainly for the US market (see Marchak, 1983; Woodbridge, Reed, 1984). Unfortunately, for a variety of reasons, including the development of more efficient ways of utilization and the creation of substitute products, market demands for these commodities is unlikely to increase by much, if at all, and will almost certainly remain volatile. The point is reinforced by the prolonged extent of recessionary conditions facing the British Columbia forest product industries compared to other regions such as Central Canada, the southern US and Sweden where there is more emphasis on technologically sophisticated higher-value products.

The philosophy that British Columbia's forest product industries serve as marginal and therefore vulnerable units on the global scene is deeply entrenched. Such a philosophy, or 'export staples mentality', helps explain both the industries' doctrinaire support for free trade, and for its strong conservatism towards research, development and innovation. Thus, the widespread (and widely noted) reliance upon adapting 'proven' technology which has been tried and tested elsewhere is typically justified in terms of not compounding the uncertainties of exporting or, if appropriate, of locating in a remote region. Concentration on a few bulk commodities has also pre-empted marketing initiatives; such commodities are simply traded. In the sawmill industry, for example, there has been a tendency to concentrate on a limited range of dimension lumber products. In addition, as Woodbridge, Reed (1984) point out, the technologically conservative attitudes of forest product firms have discouraged innovation in the backwardly linked machinery and engineering industries which are likewise 'locked-in' to maturing technologies. Moreover, a high level of foreign ownership has reinforced the related characteristics of bulk commodity specialization and technological dependence (Hayter, 1985).

The underlying problem of the British Columbia forest industries is therefore an undue specialization on a narrow range of products, markets and mature technologies. As Marchak (1983), Woodbridge, Reed (1984) and others (e.g. Hayter, 1981 and 1982) have noted, an increased commitment to technological innovation is essential if the industry is to escape the confines of a marginal producer. Although difficult to state precisely, technological demands and opportunities are clearly huge. It is likely, for example, that modernization of the British Columbia forest product industries designed to achieve contemporary state of the art production systems would require about $4b worth of investment. Certainly, the past decade or so has been a period of considerable technological change in the forest product industries in general. In particular, there have been major innovations in logging, pulping, paper making, paper finishing, bleaching, and wood products as well as extensive application of microelectronics to all aspects of forest product manufacturing operations (Hayter, 1986). All the signs indicate that techno-logical change will continue to be important and that technological head starts, no matter how short, will be a source of competitive advantage.

It is generally conceded by industry that innovation is important to improve labour productivity. Technological innovation is also necessary to utilize more effectively a changing and still imperfectly understood resource base, to develop and adapt technologies for local conditions and, most importantly, to penetrate in a faster more flexible way a much wider range of industrial and geographical market segments, especially higher-value market segments. In this latter regard, for example, a shift to a more diverse and higher-value product mix can be achieved by innovating technology which upgrades existing products, adds processing steps, and develops new or modified products (Woodbridge, Reed, 1984, p. 11). Upgrading possibilities, for example, are particularly evident in the lumber industry through species separation, finger-jointing and machine stress rating. Additional processing steps could be realized by the establishment of remanufacturing facilities adjacent to sawmills and by the production of various kinds of high-value coated and uncoated papers at existing pulp and paper mills. It should be stressed that at present in British Columbia the manufacture of wood-based consumer products is extremely limited. The most popular source of a very wide range of medium-priced wooden furniture in Vancouver, for example, is IKEA, a Swedish-owned store which imports its products ready to assemble.

In the wood-processing industry there is also potential for the development of new products such as specialized boards and panels manufactured from 'waste' material. In addition, it is evident that the Japanese construction industry offers a potentially huge market *if* British Columbia producers show greater sensitivity to the specialized needs of that market. In this regard, greater flexibility in production processes, to allow the seeking out of specific market niches, would be facilitated if production decisions were based on maximizing the value of log conversion, rather than the quantity of log conversion which is the current norm. It is also apparent that, even in such a traditionally capital-intensive activity as pulp manufacture, and as Piore and Sabel (1985) reported for other mature industries, viable small-scale operations in British Columbia are technologically feasible. Technological choices regarding products and processes are therefore available to the British Columbia forest industries.

The nature of the technological challenge facing the forest industries, bearing in mind their size and scope within the provincial economy, strongly suggests that they be made a priority in any policy to stimulate innovation. In this regard, the general thrust of this policy would require more investment in in-house R & D, closer liaisons between universities, government laboratories and industry, a stronger marketing effort by industry and, perhaps most important of all, the adoption of more innovative attitudes. In addition to maintaining the vitality of the province's most important industries, an innovation policy for the forest industries could be used as the basis for promoting and focusing involvement in such key high technology areas as microelectronics, robotics and biotechnology. Moreover, investments in value-added facilities and R & D groups, and any stimulus to local engineering services and equipment manufacturers, would be employment enhancing and would offset the job losses associated with productivity improvements. If the government can also introduce an effective forest management strategy, employment could even exceed the peak 1979 levels. The barriers to success along these lines should not be underestimated. Nevertheless there are signs that some firms are recognizing the need to become more innovative which, in association with recent Canadianization of the sector, at least gives cause for hope.

INDUSTRIAL POLICY, INNOVATION POLICY AND MATURE INDUSTRIES: CONCLUDING COMMENT

Industrial policy refers to the activities of governments which are designed to influence the global competitiveness of industries in a national (regional) economy. Industrial policies may be explicit or implicit, positive or negative, *ad hoc* and contradictory or comprehensive and cohesive. It is important, however, to stress the contingent nature of industrial policy. That is, an appropriate policy for an economy may change over time and differ from the policy appropriate to other economies. In this regard, *modern* innovation policy represents a fusion of traditional science and technology policy and traditional industrial policy (Rothwell and Zegweld, 1981, p. 1). In general, innovation policy interprets comparative advantage in more dynamic terms, in which the international division of labour is shaped by the planning, organization, education and creative power of the work force and of institutions and by their choices and abilities to adapt to changing economic circumstances. In practice, innovation policy has come to be associated very narrowly with 'high tech' activities, while debates regarding mature industries continue to be couched in the rhetoric of the traditional (static) concepts of comparative advantage (Reich, 1982). Particularly in the case of hinterland regions, however, a strong case can be made for giving more attention to mature industries within the context of innovation policy.

First, as conventional innovation theory itself indicates, productivity-oriented innovations will normally be important for continued competitiveness and there may be circumstances when the rate of innovation falls unacceptably low or an industry loses its ability to adapt technology developed elsewhere to its own circumstances. This latter point, of course, is of particular importance in the context of resource industries. Second, a greater commitment to innovation may be essential to creating value-added activities which even in mature industries can be job enhancing. Third, mature industries are typically strongly connected to local supplying industries and services so that innovations by the former will have wider repercussions on the regional economy. Fourth, mature industries may themselves consist of a range of low, medium *and* high tech businesses. That is, emerging research intensive businesses are found in segments of older industries. In addition, it is possible that mature industries may provide a seed bed and orientation in which to develop expertise in the newly emerging technologies. Fifth, as Mansfield et al., (1977, chapter 9) point out, the rates of return on R & D on low tech activities can be surprisingly high. Indeed, these authors suggest that too much emphasis is being placed on high tech activities. Sixth, it may be argued that the adaptation required of the work force, for example, in the form of retraining, as a consequence (or expectation) of innovation policy is more likely to be successful if workers can remain within their existing industrial environment.

The last point to be made in this context is that science and technology underlies production change in all goods-producing industries and not just high tech activities. An innovation policy emphasis on mature industries would typically be able to draw upon, as well as contribute to, accumulated investments in physical and human resources *and* in a manner which could favour hinterland economies. In the specific case of British Columbia, for example, the ambitions of the provincial government regarding innovation policy may be best met by seeking to establish a leading edge position in forest product technology. From a regional and national (and international) perspective such a thrust has much to commend it, possibly more so than the dream of becoming another Silicon Valley clone.

ACKNOWLEDGEMENTS

The author would like to thank SSHRC for a travel grant to the symposium at which this chapter was presented as a paper.

REFERENCES

Abernathy, W. J. and Utterbach, J. M. (1978): Patterns of industrial innovation, *Technology Review* 81, June–July, 40–47.

Abernathy, W. J., Clark, K. B. and Kantrow, A. M. (1983) *Industrial Renaissance: Producing a Competitive Future for America*, Basic Books, New York.

Britton, J. N. H. and Gilmour, J. M. (1978) *The Weakest Link: a Technological Perspective on Canadian Industrial Underdevelopment*, Background Study 43, Science Council of Canada, Ottawa.

Cohen, D. and Shannon, K. (1984) *The Next Canadian Economy*, Eden Press, Montreal.

Ewers, H. J. and Wettmann, R. W. (1980) Innovation oriented regional policy, *Regional Studies* 14, 161–80.

Farris, L. D. (1983) *The B.C. Forest Industry to 1990*, Canada Employment and Immigration Commission, Vancouver.

Freeman, C. (1978) Preface. In *Policies for the Stimulation of Industrial Innovation* Analytical Report, Vol. 1, OECD, Paris, 5–14.

Freeman, C., Clark, J. and Soete, L. (1982) *Unemployment and Technical Innovation: a Study of Long Waves and Economic Development*, Frances Pinter, London.

Gaudry, R. (1976) *The State of Research and Research Funding in British Columbia*, Queen's Printer, Victoria.

Hayter, R. (1976) Corporate strategies and industrial change in the Canadian forest product industries, *Geographical Review* 66, 209–28.

Hayter, R. (1981) Patterns of entry and the role of foreign-controlled investments in the forest product sector of British Columbia, *Tijdschrift voor Economische en Sociale Geografie* 72, 99–113.

Hayter, R. (1982) Research and development in the Canadian forest product sector-another weak link, *Canadian Geographer* 26, 256–63.

Hayter, R. (1985) The evolution and structure of the Canadian forest product sector: an assessment of the role of foreign ownership and control, *Fennia* 163, 439–50.

Hayter, R. (1986) *Technology Policy Perspectives and the Canadian Forest Product Industries* (prepared for the Science Council of Canada, Ottawa).

Hayter, R. and Gunton, T. (1984) British Columbia's Discovery Park Policy: a regional planning perspective. In Waters N. M. (Ed.), *Nanaimo Papers*, Western Division, Canadian Association of Geographers, Tantalus, Vancouver, 27–42.

Hirsch, S. (1967) *Location of Industry and International Competitiveness*, Clarendon Press, Oxford.

Hollander, S. (1965) *The Sources of Increased Efficiency: a Study of DuPont Rayon Plants*, MIT Press, Cambridge, Mass.

Joint Economic Committee, Congress of the United States (1982) *Location of High Technology Firms and Regional Economic Development*, US Government Printing Office, Washington.

Mansfield, E., Rapoport, J., Romeo, A., Villani, E., Wagner, S. and Husic, F. (1977) *The Production and Application of New Industrial Technology*, W. W. Norton & Co, New York.

Marchak, P. (1983) *Greengold: the Forest Industry in British Columbia*, University of British Columbia Press, Vancouver.

Mensch, G. (1979) *Stalemate in Technology*, Bulinger, Cambridge, Mass.

Nelson, R. R. (1984) *High-Technology Policies: a Five Nation Comparison*, American Enterprise Institute for Public Policy Research, Washington.

Piore, M. J. and Sabel, C. F. (1985) *The Second Industrial Divide*, Basic Books, New York.

Reed, F. L. C. and Associates Ltd. (1975) *The British Columbia Forest Industry: its Direct and Indirect Impact on the Economy*, Vancouver.

Rees, J., Briggs, R. and Hicks, D. (1985) New technology in the United States machinery industry: trends and implications. In Thwaites, A. T. and Oakey, R. P. (Eds.) *The Regional Economic Impact of Technological Change*, Frances Pinter, London, 164–94.

Reich, R. (1982) Beyond Free Trade, *Foreign Affairs* 60, 852–81.

Rothwell, R. and Zegweld, W. (1981) *Industrial Innovation and Public Policy*, Frances Pinter, London.

Schott, K. (1981) *Industrial Innovation in the United Kingdom, Canada and the United States*, British-North American Committee, London.

232   *Hayter*

Steed, G. P. F. and DeGenova, D. (1983) Ottawa's technology oriented complex, *Canadian Geographer* 27, 263–78.

Thomas, M. D. T. (1985) Regional economic development and the role of innovation and technological change. In Thwaites, A. T. and Oakey, R. P. (Eds.) *The Regional Economic Impact of Technological Change*, Frances Pinter, London, 13–35.

Utterbach, J. M. and Abernathy, W. J. (1975) A dynamic model of process and product innovation, *Omega* 3, 639–56.

Vernon, R. (1966) International investment and international trade in the product life cycle, *Quarterly Journal of Economics* 80, 190–207.

Vos, D. (1983) *Governments and microelectronics*, Science Council of Canada, Ottawa.

Watkins, M. H. (1963) A staple theory of economic growth, *The Canadian Journal of Economics and Political Science* 29, 141–58.

Wells, L. T. (1968) A product life cycle model for international trade, *Journal of Marketing* 32, 1–16.

Whittington, D. (Ed.) (1985) *High hopes for high tec*, University of North Carolina Press, Chapel Hill.

Woodbridge, Reed and Associates (1984) *British Columbia's Forest Products Industry: constraints to Growth*, prepared for the Ministry of State for Economic and Regional Development, Vancouver.

12

Promoting 'High Technology' industry: location factors and public policy

C. V. Christy and R. G. Ironside

Silicon Valley, the legendary California land where the microprocessor was born, has produced a host of imitators: Silicon Prairie (the Dallas-Fort Worth area), Silicon Glen (Central Scotland), Software Valley (the Thames-Kennet Valley from Maidenhead to Newbury), and doubtless almost as many others as there are technological journalists. But it has produced not only names; it has brought forth a new economic Holy Grail, industrial renaissance through high technology job creation. Every city in the advanced industrial world, it now seems, is struggling to open its science park as the answer to decaying steel mills and rusting automobile plants (Hall and Markusen, 1985, p. vii).

Each era has impoverished and growth regions, technological advances and declines. While government regional development policies have moderated regional disparities, these disparities have persisted. International trade competition, particularly from Asia, has reduced the markets for old products from low technology industries of Western Europe and North America. That competition, coupled with the effects of two recessions since 1974, forced attention away from regional problems to national ones requiring manufacturing industry to be restructured. In this context high technology has acted as a beacon for governments. The microchip, laser, robotics, computer graphics, telecommunication and semiconductor industries are widely perceived to offer a longer term solution than the 'quick fix' of branch plant jobs or subsidized employment in 'sunset' industries. This chapter reviews recent experiences in the public promotion of 'high technology' industry, paying particular attention to the factors which seem to affect the location of such activities. It proceeds from a general discussion of these issues to a specific analysis of the nature of high technology industry in Alberta, in the context of the provincial government's efforts to diversify what is essentially a resource-based peripheral economy.

Despite their intuitive appeal, policies to promote 'high technology' are often based upon a misunderstanding of the role of technical change in economic development. In particular, they frequently fail to make the distinction, noted by Freeman and McArthur in the first two chapters of this book, between those technologies which may be expected to generate growth in the medium term and those likely to produce a more immediate economic impact. On a more pragmatic level, public agencies vary widely in defining the 'high technology' targets for their policies. For example, occupational characteristics, such as the number of highly qualified technical and scientific staff are frequently used as criteria (Langridge, 1984). In contrast, the Connecticut High Technology Council focuses on 'any industry that is going to create jobs in the 1980s and 1990s' (Payne, 1983). Glasmeier et al. (1983a) note a misconception in the literature in that 'high technology' industries are treated as a homogenous group. Browne (1983) suggests that 'high technology' cuts across established industrial classifications. Thus, high-technology processes are being adopted by low-technology industries such as automobile assembly whilst, at the same time, Markusen (1983) perceives that two thirds of jobs in the US computer software industry are as low skilled as in low-technology industry.

There are, nevertheless, some outstanding features of 'high-technology' industries: they have a high proportion of small, new independent firms whose hallmark is their R&D and innovation of new products.[1] Frequently they grow rapidly and are perceived as incipient major firms. Steed (1982) labelled those medium-sized firms (100–2499 employees) which had grown rapidly in high-technology-intensive sectors, 'threshold firms', with potential for significant expansion. It is this potential which makes such firms attractive to policy makers, and the identification of the circumstances which have led to their agglomeration in growth complexes such as Silicon Valley, Route 128 or the Research Triangle is obviously important in attempts to replicate these conditions elsewhere.

LOCATIONAL DETERMINANTS OF HIGH-TECHNOLOGY FIRMS

Table 12.1 shows locational factors identified by the few major studies which have reviewed high-technology firms. Most of these factors have been ingredients in the emergence of the previously mentioned agglomerations. Underpinning the case of Silicon Valley was a clear symbiosis between private

Table 12.1 Factors that influence regional location i.e. preferences of high technology companies: summary of locational surveys

Joint Economic Congress Survey (1982)		Congress of the United States office of Technology Assessment Survey (1984)		Glasmeier et al Survey (1982)		Bednarz's Texas Locational Study (1984)	
Rank	Factor	Rank	Factor	Rank	Factor	Rank	Factor
1	Labour skills/availability	1	Founding entrepreneurs lived there	1	Labour force	1	Proximity to airports
2	Labour costs	2	Close to existing operations	2	Airports	2	Good highway access
3	Tax climate	3	Labour skills/availability	3	Defense spending	3	Room for future expansion
4	Academic institutions	4	State government support	4	Universities	4	Professional and technical labour
5	Cost of living	5	Local transportation		Netherlands Locational Study (in Bollinger et al. 1980)		5	Labour costs
6	Transportation	6	Quality of life	1	Shortage of skilled labour	6	Proximity to large urban centres
7	Access to markets	7	High technology business climate	2	Increased competition		
8	Regional regulatory practices	7*	Universities	3	Lack of venture capital		
9	Energy cost/availability	9	Availability of suitable sites				
10	Cultural amenities	10	Overall business climate				
11	Climate	11	Financial incentives				
12	Access to raw materials	12	Venture capital availability				

* Tied for seventh rank.

entrepreneurs and Stanford University, and in the case of Route 128 between firms and the Massachusetts Institute of Technology and Harvard University. Where the cases differ according to Dorfman (1983) and the Joint Economic Congress of the United States study (1982), is that the universities associated with Route 128 were not as directly involved in the development as Stanford University was through the academic entrepreneurship of Professor Frederick Terman. He initiated research in electrical engineering into high technology and the idea for a research park on Stanford's campus in 1951. Today, some fifty one parks exist in Silicon Valley, and it is reported by Dorfman (1983) that sixteen were developed along Route 128 by the late 1970s. In both cases, academic programmes were developed in the universities and graduate students and staff created new firms. The network of information was important then as it still is (Leonard-Barton, 1984). Substantial numbers of new firms were also 'spun' from original ones, as in the case of Fairchild alumni in Silicon Valley (Harrington, 1985), while subsequent mergers and acquisitions brought new capital to the area.

Another major influence was the propulsive effect of government procurement for defence during and since the Second World War and also for the space programme. By 1982, 30 per cent of the income of Route 128 firms was generated by defence contracts (Joint Economic Congress Study, 1982). Subsidiaries of major corporations carrying out R&D also benefited. State and municipal government investments in capital infrastructure facilitating growth should also not be forgotten. There were two other advantages. Between 1970 and 1980, California and Massachusetts were the only two states consistently attracting venture capital (Congress of the United States Office of Technology Assessment, 1984). The universities associated with these developments were also significant leaders in their fields: Stanford in electrical engineering, Massachusetts Institute of Technology (MIT) in computing science and Harvard in business education.

In contrast to these cases, the Research Triangle Park in North Carolina was founded in 1959 by the State government to provide employment for graduates from three adjacent universities: North Carolina State, University of North Carolina and Duke University. The Research Triangle Foundation, a non-profit research affiliate to the universities and the state-sponsored North Carolina Science and Technology Research Centre, the first of its kind in North America, were also created. A 'hard-sell' promotion attracted branch plants of major corporations such as IBM which located research and manufacturing facilities in the park in 1965. The Joint Economic Congress of the United States study (1982) summarized the factors contributing to the success of the Research Triangle Park as follows: a critical mass of educated and technical workers; competitive low labour costs and labour availability; park location; quality of life in the North Carolina 'sun belt' and integrated promotion efforts.

Distilling the experiences of these three high-technology agglomerations and the results of major studies (table 12.1), it is clear that there is no agreement on the ranking of locational variables, although labour-related ones all rank highly. Individual influences differ in their significance between locations, firms and industries and a combination of locational influences contribute to the developments in each case. Some *caveats* have to be noted, therefore, in establishing the importance of location variables. They are:

1 *Universities do not necessarily attract high technology firms.* In 'Silicon Valley North' around Ottawa, federal government research agencies such as the National Research Councils' laboratories, the private laboratories of Bell Northern Research Ltd and Computing Services of Canada, a defence-related firm, were important locational attractions, not the Universities of Ottawa or Carleton. Federal purchasing contracts were significant as well as the information resources of the capital region and skilled labour (Steed and DeGenova, 1983). Only recently has the Research Centre for High Technology Management been established at Carleton University. Similarly in England and Scotland it has been only recently that significant interaction between high technology firms and universities has occurred. Research parks with strong university ties have emerged at Edinburgh, Warwick and Cambridge. The Cambridge area now has some 350 firms employing 14 000. But Oakey (1984) discovered that in the Silicon Valley San Francisco Bay region, independent instrument and electronic firms had few links with universities.

2 *A research park location is not indispensable for high-technology development.* By 1967 there were 126 in the USA and Canada with an occupancy rate of 27 per cent! (Danilov, 1967, 1971). Danilov concluded that parks had a 25 per cent success rate in the US. A recent study indicated there were 81 research parks in North America with an average occupancy rate as low as 37 per cent (Hayter and Gunton, 1983). In Canada ten parks exist, eight created in Western Canada since 1980, all by provincial governments. All the western parks have few or no tenants although it is too early to conclude they are failures. Parks have not caused high-technology development in Britain, the Netherlands or West Germany, though they have facilitated it (Lowe, 1985).

3 *While small entrepreneur-founded, independent firms, are a propulsive innovative force for any high-technology development, subsidiaries of large*

corporations carry out *R&D and stimulate high technology linkages*.
Numerous examples include firms in the Triangle Park and R&D in
US subsidiaries in Scotland.

4 *A strong public expenditure component* is found in all developments either
through creation of markets via defence expenditures, research
funding, education programmes, aid to improve business climate or
capital infrastructure.

5 Malecki (1984) and others have stressed *access to recreational and
cultural amenities of metropolitan cities* as being necessary to attract
scientists, engineers and managers. However, Glasmeier et al. (1983b)
suggest this locational factor has been overstated. Not only are large
urban areas in the US not attracting new jobs in high-technology industries, but nine out of the top ten metropolitan areas in percentage of
labour force in high technology industry are small to medium-sized
centres. A decentralization of the industry is occurring at least in some
products at late stages in the short product cycle. Hall and Markusen
(1985) indicate that decentralization, including overseas locations is
occurring in software marketing, servicing, semiconductors and computers. The high cost of living in metropolitan cities, the high cost of
US labour and the need to follow certain markets are all important.
Glasmeier et al. (1983b) discovered that defence, aerospace, semiconductor and resource-based high-technology firms were highly
concentrated, while computing and firms associated with mature
producer goods industry such as chemicals and fertilizers were more
dispersed because of the reasons just given.

Nevertheless, Hall and Markusen (1985) have pointed out that agglomeration
economies are strong in existing high-technology regions. It is unlikely that
major decentralization will occur as long as product innovation continues which
is, by definition, necessary in the competitive defence, space and consumer
durable manufacturing sectors. The locational 'glue' of a skilled professional
and technical labour supply, specialized contractors, venture capital, information resources, good R&D facilities in government or university facilities,
the 'proper address' as well as the attractions of cities will bind together the
high-technology firms in the existing agglomerations. Recently, Segal (1985)
described the 'Cambridge phenomenon' where such a critical mass of
interactive firms and entrepreneurs developed, subsequently attracting
support services and venture capital. At late stages in the product cycle for
reasons of market penetration and labour cost, firms may locate, however,
in peripheral regions, and other countries. However, peripheral locations
are viewed generally as being disadvantaged for firms with new products

because they lack the levels of entrepreneurship, support services and capital present in city locations (Congress of the United States Office of Technology Assessment, 1984). Nevertheless, as in Texas and Colorado, leading innovative high-technology firms can be enticed, sometimes by major funding, to attractive though isolated cities, such as the case of the Microelectronics and Computer Technology Corporation (a consortium of firms), which relocated from the east coast to Austin, Texas.

Turning to the results of studies reviewing locational variables (table 12.1), the most significant factors appear to be the availability of labour skills, accessibility and business climate. The Joint Economic Congress study (1982) found labour skills, costs and tax climate to be the most important factors. Glasmeier et al. (1983b) cite labour force, proximity to airports and defence spending. The study by Bednarz (1984) in Texas conversely placed accessibility variables and site space ahead of labour skills and costs. The United States Office of Technology Assessment survey (1984) of 99 high technology firms ranked, in order of importance; home of the founding entrepreneur, closeness to existing operations, a skilled labour force. Proximity to universities co-ranked seventh.

The last-mentioned study was the only major survey to identify the significance of the original place of residence of the entrepreneur. Yet in Silicon Valley, Cooper (1970) discovered that in 97 per cent of the firms surveyed, at least one founding entrepreneur resided locally prior to incorporation. Oakey, Thwaites and Nash (1980) found that in the majority of cases in Britain, the first commercial production of 323 significant product innovations was in the same region as the inventor's firm. In the Ottawa region, research by Steed and DeGenova (1983) indicated that 41 of 45 firms were founded by entrepreneurs with strong local associations.

GOVERNMENT PROMOTION OF HIGH-TECHNOLOGY INDUSTRY

Given the diverse locational and other characteristics of high-technology firms, governments are faced with difficulties in selecting the most effective incentives to encourage indigenous entrepreneurs or to attract firms from elsewhere. This has not deterred many from trying. The United States Office of Technology (1984) study, for example, revealed that twenty two states had some form of development programme for high technology, but many were not distinct from programmes designed to attract any foot-loose industry. A wide range of public policies has been adopted at national and local level, but, for most governments, the model of 250 000 employed in the 1800 firms of Silicon Valley will prove to be a mirage.

Research Parks

The initial success of privately developed examples in Silicon Valley and along Route 128 encouraged replication by governments. The concept links an academic or research institution with the business community. Investment in basic and applied research is undertaken with an emphasis on technology transfer into innovative commercial products. Generally, manufacturing, sales offices and warehouses are excluded from research parks. Manufacturing facilities have developed, however, in close proximity to such parks in the major American high-technology areas. The proliferation of speculative private and public research parks in North America, as just mentioned, has saturated the market, resulting in low occupancy rates.

Innovation Centres

They were introduced in the 1970s at the US universities of MIT, Carnegie-Mellon, Oregon, Utah and in Canada at Waterloo and the École Polytechnique de Montréal. Links between university and industry in applied research, technology transfer, entrepreneurship programmes and innovation are objectives. There are now more than 400 North American academic institutions offering entrepreneurship courses (Congress of the United States Office of Technology Assessment, 1984). Often, incubator premises are provided in such centres. Success is recorded at Waterloo University where 19 firms were created between 1980–3 (Walker, 1983) and the Ontario government will spend $100 million establishing a network of centres. European examples which exist in Britain and the Netherlands have stimulated the development of some 70 centres in West Germany (Meyer-Krahmer, 1985).

Public Venture Funds

While new firms in North America and Western Europe have relied on personal savings of founding members and private venture capital for start up and growth, such venture capital sources have been few in Western Europe. Since the late 1970s, however, public venture funds have been created by governments. Britain, although now a European leader in public venture capital through the British Technology Group and the loan guarantee and the business expansion schemes of the Government, can trace such public investment in advanced technology back to 1945 (Rothwell, 1985). The two British schemes just mentioned introduced since 1981, have played a large role in also developing the supply of private venture capital. Elsewhere, for example, the Deutsche Waginisfinanzierungeschaft (The German Venture Finance Company) was established by the federal government of West Germany in 1975. Most public funds do not seek majority ownership or management of the firms. However, some US state governments are placing explicit geographical requirements on firms to ensure within-state development. In Canada, only one public fund exists, Vencap Equities Ltd, which was created by the Alberta Government in 1983. The public participates through the purchase of shares. So far its investments have proved to be few and conservative. Other sources of public venture capital are available, however, in each province.

Subsidies

These encompass grants and tax incentives specifically directed at R&D, new or small firm development. They can be varied in size and timing and linked to the eligibility of firms, industrial sector and location. Their effectiveness is debatable. In Canada the Federal Task Force on Federal Policies and Programs for Technology Development (Ministry of State Science and Technology Canada, 1984), found that such programmes did not assist new firms but rather well-established ones. Similar findings have been made in Western Europe (Rothwell, 1985).

Tax incentives via credits or depreciation rates can reduce start-up costs. To what extent they promote innovation or new firm development is unclear. Recent evidence in Canada, a pioneer in R&D tax credits and allowances, suggests it is very modest (Mansfield and Switzer, 1985). However, in Britain a corporate tax reduction was found to stimulate R&D more than any other policy (Schott, 1981). In the US, in contrast, new firms find such measures useful but only as one form of government support (United States Office of Technology Assessment, 1984). Governments can also make mistakes in over generous programmes which have then had to be cancelled.

Government Procurement

Purchasing from private firms, particularly of defence and aerospace products, has stimulated high-technology firms in North America and Europe (see chapter 10). Co-ordinated purchasing as a deliberate policy of government is, however, usually lacking. It can be an unstable policy instrument because of changes in government policy and governments. In the long term it could reduce competition, cost effectiveness in public expenditures and result in purchasing errors because of bureaucratic lack of knowledge (Rothwell and Zegveld, 1981).

'Picking Winners'

Greater efforts are being devoted to the identification of innovating small to medium-sized firms which contribute to indigenous regional growth as targets for government assistance. Caution is necessary, however, if the past lack of success of governments in selecting firms in low-technology industries to be recipients for aid, is considered. Can governments duplicate the success rate of private venture capital investments? Again a successful forerunner is one reason for government enthusiasm. The collaboration of corporate, financial and government institutions through the Japanese Ministry of International Trade and Industry to promote certain industries has encouraged other central governments to believe they can foster new industrial growth as opposed to propping up 'sunset' industries. Steed (1982) identified potential winners among five technology-intensive sectors in Canada. These sectors were identified, using the two-digit SIC, by expressing research intensity in terms of R&D as a percentage of value added. They included electrical products, petroleum and coal products, machinery, chemical products and transport equipment. In total, they accounted for 75 per cent of the 1975 R&D expenditure by Canadian industry. Yet should apparently declining industries such as textiles, clothing and automobiles be ignored? The partial rejuvenation of the automobile industry in western Europe and North America has been facilitated by new technology (see chapter 7) and, in chapter 11, Hayter has provided a convincing case for the inclusion of mature industries in technology policies.

This review of policies and programmes is by no means comprehensive, but it is clear that, among the wide array available, with many permutations internationally, there are as yet no 'winners'! One of the problems is the stereotype official perception of high-technology industry as clean, small scale, employing high salaried professional and technical workers, manufacturing a high-value compact, product (Bednarz, 1984). Some industry and some firms have these characteristics, others do not. Undoubtedly, the stereotype has contributed to the inability of many governments to distinguish between high-technology industry and the 'light' industry of yesteryear and between the different types of high-technology industry and their associated locational, labour skill and financial needs. In this context, a recent study was carried out in Alberta to classify the nature of high-technology industry and, more specifically, to identify locational factors affecting its development, to establish its relationship to the oil and gas industry of the Province and to assess its dependence upon external capital relative to indigenous new firm growth.

HIGH-TECHNOLOGY DEVELOPMENT IN ALBERTA

Alberta is a classic resource-rich, manufacturing-poor, peripheral region in Canada. As such, its government aims to diversify the economy away from nonrenewable oil and gas resources and to encourage economic activities to locate to regional cities rather than only to the dominant cities of Edmonton and Calgary. High-technology industry is perceived as one means of accomplishing these objectives (Government of Alberta, 1984). Potential 'winners' exist in fibre optics, integrated circuits, computer software and biotechnology, while major research centres and universities exist. Finance is available also in the Heritage Savings Trust Fund created by the government in 1976 with revenue from oil and gas sales.

On the basis of a detailed literature review and existing knowledge of the Alberta economy, a series of research hypotheses was derived. These may be divided into two broad categories; those (hypotheses 1–3) concerned with the nature of high-technology industry in Alberta and those (hypotheses 4–8) which focus upon locational issues. The hypotheses were

1 that high technology in Alberta is associated predominantly with resource-based and service-oriented firms rather than with the stereo-type computer, semiconductor and electronics manufacturing firms.
2 that most medium-sized and large high-technology firms in Alberta are subsidiaries of foreign and Canadian firms with head offices outside Alberta as opposed to independent Alberta-owned companies.
3 that most independent firms started by entrepreneurs are young, that is less than five years old. In addition, they are small with less than 100 employees and they do not have subsidiaries.
4 that high-technology firms do not prefer research-park location
5 that labour skills and labour costs are more important location influences than proximity to a university or government research facility for transfer of research results.
6 that tax climate and provincial government assistance are important locational influences.
7 that the location of the home of the founding entrepreneur often explains location of the firm.
8 that metropolitan locations will be preferred to regional cities or large towns because of the availability of services.

In order to establish a population of high-technology firms recourse was made to a list from the Alberta Department of Economic Development. This list was ascertained to be incomplete. The lists of the Edmonton and Calgary

Research Park Authorities (Edmonton Council for Advanced Technology, 1984; Calgary Economic Development Authority, 1984), the Electronics Directory (Electronics Industry Association of Alberta, 1984) and the Nickle's Oil and Gas Directory (Southam Energy Group, 1984) were also scrutinized in order to compile the most up-to-date list in Alberta. Mindful of the definitional difficulties surrounding 'high technology', for the purposes of the study as broad a definition as possible was adopted to include firms in different industries. Using a variety of information to ascertain product, service and whether or not R&D was conducted, the total population of high-technology firms in Alberta was identified as 339 firms.[2] This number allowed, following a pre-test, a 100 per cent mailed questionnaire survey. There were 116 usable returns giving a response rate of 34.2 per cent. In terms of company officers completing responses 44 per cent of firms were completed by the Chief Executive Officer, 18 per cent by the Vice President and 23 per cent by managers.

Results and Analysis

Hypothesis 1 Using SIC categories, 28 per cent of firms described their major business activity as business management or services; 26.7 per cent were in miscellaneous manufacturing; others were in the communication industry (12.1 per cent), electrical production (10.3 per cent) and mining services (8.6 per cent). Major products or services by sales were computers (32.7 per cent), electrical and communications (30 per cent), consulting (10.9 per cent) and in the petroleum industry (9.1 per cent).

While, inevitably, a major market for the management/service/consulting firms exists in the resource industries, the responses showed that firms were not directly associated with them. Main products were strongly in the electronic, computing and communication areas. Firms providing services (38.7 per cent) were also a strong component of high technlogy development. Hypothesis 1 must, therefore, be rejected. *Hypothesis 2* Independent firms were 63.7 per cent, Canadian subsidiaries 18.6 per cent and foreign subsidiaries 17.7 per cent. In terms of size, 31 was the mean number of full-time employees for independents, 542 for Canadian subsidiaries and 380 for foreign subsidiaries. A statistically significant (5 per cent or less) cross tabulation of firm ownership by size was made. There were no independents with 1000 or more employees. Fifty eight out of 63 had less than 100 employees. Fourteen subsidiaries had less than 100 employees; fourteen between 100–999 and three firms more than 1000 employees. All the independents had an R&D programme, as did 13 out of 18 Canadian subsidiaries and half the 16 foreign subsidiaries.

Most medium to large high-technology firms in Alberta are subsidiaries and not independent firms. It is also true that as subsidiaries they conduct R&D. Hypothesis 2 can be accepted therefore.

Hypothesis 3 Most firms (64.6 per cent) were founded more than six years ago during the boom period 1973–81. Only 36.3 per cent were created during the last five years. In terms of independent firms alone, 26 were established within the last five years, 19 between six and ten years ago and 22 more than ten years ago. Most independent firms are not young. They did account for 79 per cent of high-technology firms, however, created in Alberta in the last five years. It is worth noting that 26 of 41 subsidiaries, particularly foreign subsidiaries, were established in Alberta more than ten years ago. A small majority of firms (56.5 per cent) including independents, had not established subsidiaries, but a surprisingly high proportion (43.5 per cent) had done so, generally with one subsidiary only. Major locations were Alberta (28.1 per cent), the USA (21.9 per cent) and Eastern Canada (18.8 per cent). In addition, as previously noted, most independents were small in numbers employed. The third hypothesis can be confirmed, therefore, in terms of firm size and their lack of subsidiaries, but not in terms of the age of independent firms.

Hypothesis 4 Only 17 per cent of responses considered research parks to be important locations. While responses indicated that such a location was perceived to raise a firm's profile, improve communications and stimulate ideas in a more attractive working environment, a major barrier was the cost of relocating into one of the parks. The hypothesis can be accepted.

Hypothesis 5 The supply of labour skills was regarded as very important or important by 47 per cent of firms and by 10.5 per cent as being slightly important. Twenty per cent of firms thought them unimportant. In terms of labour costs, a high 46.2 per cent of responses were neutral and only 14 per cent considered labour costs important. Thirty per cent thought them unimportant. Proximity to a university was thought to be unimportant by 40.4 per cent. Only 21.3 per cent considered it important or very important. The low response in the use of external R&D university facilities confirmed the above result. The availability of labour skills was clearly regarded as an essential factor. Labour costs, however, were not of concern. The fifth hypothesis can be accepted partially, therefore, in terms of labour skills.

Hypothesis 6 An advantageous provincial tax climate was considered important or very important by 40 per cent of respondents, but the overall business climate, which would also include federal taxes and incentives, was the highest regarded locaton factor with 77.9 per cent responding that it was important or very important (table 12.2). Provincial government support programmes were not important for 40 per cent of firms and only slightly

Table 12.2   Location factors of high-technology firms in Alberta

Rank	Factor	Very significant and Significant (%)
1	Overall business climate	77.9
2	Founding entrepreneur lived there	67.4
3	Access to Markets	60.4
4	Labour skills/availability	47.4
5	Political stability	45.8
6	Provincial tax climate	40.0
7	Local government incentives	30.6
8	Proximity to international airport	27.4
9	Proximity to domestic airport	25.3
10	Proximity to university	21.3
11	Provincial govt. support programme	21.1
12	Availability of venture capital	17.1
13	Recreational opportunities	16.8
14	Local transportation	16.0
15	Access to raw materials	15.1
16	Energy costs/availability	14.9
17	Cost of living	14.7
18	Cultural amenities	14.7
19	Labour costs	14.0
20	Proximity to Govt. depts./offices	12.6
21	Climate	6.3

Source: Firm Survey, 1985

so for another 14.7 per cent. Twenty-one per cent of firms considered them important. However, 46.8 per cent of firms did use them. Provincial tax incentives and grants for R&D were regarded as satisfactory or very satisfactory by 65.4 per cent and 53.8 per cent respectively of respondents, whilst the corresponding federal measures were even more widely welcomed. Significantly, 24.3 per cent of R&D funding was financed by the federal government, compared with 8.1 per cent from the provincial government. Overall, government influence was perceived to be very important in terms of taxes and incentives contributing to a satisfactory business climate. Political stability was considered by 45.8 per cent of firms to be also very significant or significant (table 12.2). Hypothesis six can therefore be accepted.

Hypothesis 7 The place of residence of the founding entrepreneur was an important or very important location factor for 67.3 per cent of firms (table 12.2). Hypothisis 7 can also be accepted.

Hypothesis 8 Half the firms responded that they could operate in other locations in Alberta, but 41.8 per cent could not do so. Calgary or Edmonton were the most preferred locations with towns in their commutersheds ranking next. Red Deer was the only other location which received any strong

Table 12.3 Service linkages of high technology firms in Alberta

Rank	Factor	Very significant and Significant (%)
1	Financial services	40.4
2	Engineering	37.2
3	Computer & data processing	34.0
4	Transportation	28.0
5	Government R&D laboratories	27.6
6	Office & lab rentals	24.5
7	Information banks	23.4
8	Commercial test centres	22.3
9	Legal services	21.1
10	Private R&D laboratories	19.4
11	Advertising	19.1
12	R&D parks	17.0
13	Innovation centres	15.9
14	Insurance	13.7
15	Management consulting	11.8
16	Office overload	5.3

Source: Firm Survey, 1985

preference. Supporting services, particularly financial, engineering, computer and data processing, transportation and government R&D (table 12.3) are only available to any extent in the two large cities. The metropolitan city regions dominate the locational preferences and the hypothesis can be accepted.

IMPLICATIONS FOR PUBLIC POLICY

Several conclusions with implications for public policy, may be drawn from the literature review and from the Albert study. First, it is clear that entrepreneurs are willing to invest their own capital in innovative products and independent firms. Such investment must have the prospect of excellent returns to overcome the high risks involved. Most evidence indicates that there also must be concomitantly the pull of market demand, initially at the local scale. The highest priority government policies, therefore should encourage investment by, for example, low taxes on earnings of small, new firms, R&D tax credits or grants where appropriate and procurement by government departments and universities. Cooperative R&D should also be encouraged between large firms, as is occurring in the EEC and the US, to meet international trade competition. While each of these instruments has advantages and disadvantages, the business environment can be favourably influenced by them.

The second conclusion is that high-technology firms successfully multiply in metropolitan cities where R&D and head office decision making are in close proximity and where research facilities such as government laboratories or universities exist (Oakey, 1984; Meyer-Krahmer, 1985). The city need not be large so long as there is a critical mass of creative individuals in research and business. The 'Cambridge phenomenon', Waterloo University and Ottawa's 'Silicon Valley North' are such examples. The need for higher levels of education and skills in high-technology areas is increasingly evident, although it has been pointed out that experience and business management are also necessary for success (Leonard-Barton, 1984). Of the same order of priority as the encouragement of risk investment, therefore, is the need for heavy investment by government in appropriate education and training programmes for young people and adult students in universities and colleges as well as in the work place. Because entrepreneurs tend to be from white-collar occupations (Oakey, 1984), fundamental changes through education, training and jobs have to transform the socio-economic structures of peripheral regions, traditionally blue collar in job type and industry (Wood, 1984). Another policy implication, although not of such a high priority, is that governments should also invest heavily in information and advisory services, particularly in marketing and business management. The need is especially evident in Western Europe. This would enhance the growth of regional markets and lead to eventual export markets. In Alberta the older firms are now seeking assistance from government to penetrate markets; some have established subsidiaries to do so.

The third conclusion is that while venture capital has greatly aided high technology firms, it is not always essential for their formation, although this is more true of North America than Western Europe. Personal savings and retained earnings in North America have allowed many to succeed, and the capital markets of America have underwritten the rapid incorporation and growth of firms and their subsequent mergers and acquisitions as has occurred in Silicon Valley. In Western Europe, personal capital is more difficult to accumulate and until recently private venture capital has been scarce. The pioneering regional activity of single financial institutions, however, can at times be important. Barclays Bank, for example, aided Cambridge firms and attracted other venture capital (Marshall, 1985). Government policy should be used, therefore, to increase the supply of venture capital on attractive terms from all sources both public and private. This policy area is of the same order of priority as those previously mentioned.

A fourth conclusion is that speculative private or public investment in research parks, incubator or innovation centres is not a prerequisite to develop high technology. Many firms have been successful in other locations while the cost of such premises may be affordable only as profits grow. A useful function of research parks, however, is to attract major firms which are leaders in their field in order to attract others. But the record of success of research parks has been questionable since Silicon Valley, Research Triangle and Route 128 developed. To these, the Cambridge Science Park should be added, but as Segal (1985) points out, only some 10 to 15 per cent of companies in the area are located there. Such facilities should be an object of government policy, but of a lower order of priority than others discussed.

The conclusions are generally applicable and not specific to peripheral regions. In implementing such policies, governments should provide a large premium for programmes in peripheral regions. The generally smaller scale of cities, information networks, universities, local markets, financial resources and skilled labour supply as well as the dominance of extractive resource or low-technology industries, are daunting barriers to entrepreneurs in high technology. It may be that Alberta, a Canadian peripheral region, with a large public capital fund, rich primary resources, two major cities and their universities will have sufficient ingredients to create an environment for high technology. However, the close 'networking' typical of Cambridge in the UK or Silicon Valley in California does not exist yet. This, however, is not surprising, bearing in mind the fact that approximately the same number of firms that are found in the relatively small UK city of Cambridge, are divided in Alberta between two metropolitan regions, each with a population of 600 000.

Contrary to earlier evidence, it is now argued that new high-technology firms, like most new firms, make a smaller initial contribution to overall new employment than first thought (Storey, 1983). Also, as high-growth, high-technology complexes mature, less growth may occur (Markusen, 1983). The importance of high-technology firms lies in the application of their processes and products to many industries to provide a competitive manufacturing base. While private entrepreneurs and governments may not succeed in every case if they attempt such development, if they do not try their future will be dependent on others, on primary resources and on low-technology manufacturing and service industries. In addition, a clear message from experience so far is that whatever the degree of commitment of government, the fundamental components of successful high-technology development emanate from the investment of private risk capital by entrepreneurs in relatively unplanned agglomerations of firms. In this context, public policy can act as an aid; it is not a prerequisite to high-technology development.

NOTES

1 Innovation is the initial introduction of a new product and/or the first use of a new production process. An innovation always rests upon an invention, that is on new knowledge which is transformed by the innovator into economic activity. Research and Development is generally defined as investigative and experimental work carried out to acquire new scientific and technical knowledge, to devise and develop new products and processes or to apply newly acquired knowleldge in making technically significant improvements to existing products or processes.
2 a mail survey of 50 randomly chosen nonrespondents verified the representativeness of the sample on the dependent variables of firm ownership and full-time employment size. The results indicated 15.1 per cent more independent Alberta-owned firms than the initial survey and confirmed that this group comprises the largest proportion of high-technology firms in the province. Only 8.1 per cent fewer Canadian and 4.4 per cent fewer foreign subsidiaries occurred. On full-time employment the mean was 30 per independent firm in the original survey and 31 among non-respondents firms. However, a substantial discrepancy existed between mean employment for subsidiaries. The reason lies in that they are few in number and range from 4 to 4600 employees. The majority in both groups had, however, less than 100 full-time employees.

REFERENCES

Bednarz, R. (1984) *Is High Technology Industry the Answer*, Report the Office of the Governor, Austin, Texas.
Bollinger, L., Hope K. and Utterback, J. M. (1983) A review of literature and hypotheses on new technology firms, *Research Policy* 12, 1–14.
Browne, L. E. (1983) Can High Tech save the Great Lake States?, *New England Economic Review*, Federal Reserve Bank of Boston, 20.
Calgary Economic Development Authority and Calgary Research and Development Park Authority (1984) *Advanced Technology in Calgary: Directory of Advanced Technology Firms in Calgary,*, Calgary Economic Development Authority.
Congress of the United States Office of Technology Assessment (1984) *Technology Innovation and Regional Economic Development*, Background Paper 2, US Government Printing Office, Washington.
Cooper, A. C. (1970) The Palo Alto experience, *Industrial Research*, Industrial Research Inc., Beverley Shores, Ind., May, 58–60.
Danilov, V. (1967) How successful are science parks? *Industrial Research*, Industrial Research Inc., Beverley Shores, Indiana, May, 76–82.
Danilov, V. (1971) Research park shake-out, *Industrial Research*, Industrial Research Inc., Beverley Shores, Indiana, May, 44–7.
Dorfman, N. (1983) Route 128: The development of a regional high technology economy, *Research Policy* 12, 299–316.

Edmonton Council for Advanced Technology (ECAT) (1984) *Member List*, Edmonton Research and Development Park Authority.

Electronics Industry Association of Alberta (1984) *Electronics in Alberta: A Directory of Manufacturers*, Edmonton, Alberta.

Glasmeier, A. K., Markusen, A. R. and Hall, P. G. (1983a) *Defining High Technology Industries*, Institute of Urban and Regional Development Working Paper 407, University of California at Berkeley.

Glasmeier, A. K. Markusen, A. R. and Hall, P. G. (1983b), *Recent Evidence on High Technology Industries, Spatial Tendencies: A Preliminary Investigation*, Institute of Urban and Regional Development, Working Paper 417, University of California at Berkeley.

Government of Alberta (1984) *Proposals for an Industrial and Science Strategy for Albertans 1985–1990*, White Paper, Edmonton.

Hall, P. G. and Markusen, A. R. (Eds.) (1985) *Silicon Landscapes*, Allen and Unwin, Boston.

Harrington, J. W. Jr. (1985) Intraindustry structural change and location change: US semi conductor manufacturing 1958–1980, *Regional Studies* 19, 343–52.

Haug,R., Hood, H. and Young, S. (1983) R&D Intensity in the affiliates of US owned electronics manufacturing in Scotland, *Regional Studies* 17, 383–92.

Hayter, R. and Gunton, T. (1983) British Columbia's Discovery Park Policy: A Regional Planning Perspective. In Waters, N. M. (Ed.), *Nanaimo Papers*, Western Division, Canadian Association of Geographers, Tantulus, Vancouver, 27–42.

Joint Economic Congress of the United States (1982) *Location of High Technology Firms and Regional Economic Development*, US Government Printing Office, Washington.

Langridge, R. (1984) *Defining high technology for locational analysis*. Discussion Paper in Urban and Regional Economics, Series C, No. 22, Department of Economics, University of Reading.

Leonard-Barton, L. (1984) International communication patterns among Swedish and Boston-area entrepreneurs, *Research Policy* 13, 101–14.

Lowe, J. (1985) Science parks in the UK, *Lloyds Bank Review* 156, 31–42.

Malecki, E. J. (1984) R&D and US regional development. Unpublished paper, Policy Roundtable on Industrial Geography and Public Policy, Annual Meeting of the Association of American Geographers, Washington D.C., April.

Mansfield, E. and Switzer, L. (1985) The effects of R&D tax credits and allowances in Canada, *Research Policy* 14, 97–108.

Markusen, A. R. (1983) *High-Tech Jobs, Markets and Economic Development Prospects*, Institute of Urban and Regional Development working Paper 403, University of California at Berkeley.

Marshall M. (1985) Technological change and local economic strategy in the West Midlands, *Regional Studies* 19, 571–8.

Meyer-Krahmer, F. (1985) Innovation Behaviour and regional indigenous potential, *Regional Studies* 19, 523–34.

Oakey, R. P. (1984) *High Technology Small Firms: Innovation and Regional Development in Britain and the United States*, Frances Pinter, London.

Oakey, R. P. Thwaites, A. T. and Nash, P. A. (1980) The regional distribution of innovative manufacturing establishments in Britain, *Regional Studies* 14, 235–54.

Payne, R. A. (1983) *High Technology In Canada*, Evans Research Corporation, Toronto.

Rothwell, R. (1985) Venture finance, small firms and public policy in the UK *Research Policy* 14, 253–66.

Rothwell, R. and Zegveld, W. (1981) *Industrial Innovation and Public Policy*, Frances Pinter, London.

Schott, K. (1981) *Industrial Innovation in the United Kingdom, Canada and the United States*, British-North American Committee, London.

Segal, N. S. (1985) The Cambridge phenomenon, *Regional Studies* 19, 563–70.

Southam Energy Group (1984) *Nickle's Canadian Oil Register 1983–84*, Southern Communication Ltd., Toronto.

Steed, G. (1982) *Threshold Firms: Backing Canada's Winners*, Science Council of Canada, Ottawa.

Steed, G. and DeGenova, D. (1983) Ottawa's Technology-Oriented Complex, *Canadian Geographer* 27, 262–78.

Walker, D. F. (1983) Innovation Centres and research parks as an element of regional renewal with special reference to Waterloo. In Barr, B. M. and Waters, N. M. (Eds.) *Regional Diversification and Structural Change*, BC Geographical Series, 39 Tantalus Research, Vancouver, 232–42.

Wood, P. A. (1984) Regional industrial development, *Area* 16, 281–89.

13

Conclusion

G. Humphrys and K. Chapman

There can be no denying the importance of the issues considered in this volume; it is apparent that technical change has played a central role in recent economic events which have radically altered the geography of firms and industries as well as of communities and regions. The three sections into which the book is divided represent specific aspects of this overall theme which may be related to the effective social and political management of technical change. Thus an awareness of the nature and spatial dimensions of technical change is relevant to an understanding of the processes at work, which is needed if these are to be influenced in accordance with welfare objectives. Studies of the role of technical change in the restructuring of 'mature' industries, as exemplified by chapters 6, 7 and 8, provide information about what is happening and, by inference, what needs to be done. The contributions in the final section, which focuses on some of the policy implications, are a logical progression from those in the first two, in the sense that they move from a consideration of processes and problems to the evaluation of policy and the identification of appropriate courses of action. In seeking to draw this material together, this conclusion will not attempt a comprehensive review of current thinking and literature on technical change. Such a task has already been undertaken by many with better qualifications than the editors of this volume (see Freeman and Soete, 1986; Malecki, 1983). The objective is much more modest – to identify some of the lines of further enquiry suggested by the preceding chapters.

Schumpeter's characterization of technical change as a 'gale of creative destruction' sweeping through industry is a not inappropriate description of what has been happening within academic disciplines as the extent and rapidity of technical change in recent years has exposed the inadequacies of traditional approaches. The various contributions to this volume illustrate some of the ways in which geographers have responded to this challenge.

Chapters 2 to 5 provide insights into the nature and extent of technical change which are prerequisites for the formulation of effective policies. These chapters also provide clear directions for further geographical work. Chapter two suggests that not all technical change is radical or revolutionary in character, yet little is known about the spatial impact of incremental innovation. It also points out that the institutions available play a vital role in the practical responses to technical change by public bodies of all kinds. They are not least important in the field of industrial policy and related regional development planning. Despite this there is little investigation of the kinds of institutions needed to improve policy implementation in the circumstances created by recent technical change. In the past there has been much criticism of policies and institutions by geographers and others, and few positive suggestions regarding improved alternatives. The sweeping changes accompanying the impact of information technology in particular makes it timely to undertake this task.

One criticism of some of the writings in industrial geography is that it is too willing to accept broad generalizations and models without their having been adequately tested. A counter to this argument would be more studies of the kind provided by McArthur in Chapter 3. If his results are confirmed elsewhere, the simple models of the way process and product innovation occurs will have to be modified to reflect the complexities of how this happens in reality. In one sense Wood, in chapter 4, extends this idea further, by focusing upon the significance of producer services for industry. This study reveals the need to see manufacturing production as only one element of a system which is a long chain extending from the raw material to the customer. The implication is that manufacturing production may be only a minor element in this system and so subordinate to one or more of the other elements. Technological change in one of these other elements, such as in producer services, may have more locational impact than changes in manufacturing itself. This is one of the more active areas of research at present and the results of further investigation of producer services at the national level, such as in this chapter, and at the local level, are likely to become available in the near future.

The growing interest in the impact of technical change in industry on the quantity, quality and spatial distribution of employment opportunities is reflected in Harrison's contribution (chapter 5). Technical change results in both gainers and losers in employment terms. More work is needed on the socio-economic consequences of technical change, not only in regions dominated by traditional industries such as shipbuilding, but also in those associated with concentrations of high-technology activities. These consequences are apparent in the changing character of work experiences for different sections of the labour force. Thus technical change may increase the level of job satisfaction for some, but for many others it may have the reverse effect. To the extent that these social dimensions of technical change frequently have a spatial dimension, they fall within the scope of geographical enquiry.

Pinder and Hussain open section two on technical change and restructuring with a demonstration of the way in which technical change can evoke different responses at different levels within enterprises and as between the short term and the long term. Similar studies of other major industries would provide further substantiation of their work. This would also open up the prospect of applying the results for each industry to a series of areas in turn, to expand understanding of the dynamics of regional industrial systems, which is still an elusive subject. In many ways, a systems perspective is a common element of the contributions of both Holmes (chapter 7) and Bradbury (chapter 8). They attempt to place technical change within the car and steel industries respectively in the broader context provided by Marxist interpretations of the restructuring of capitalist economies. Holmes suggests that the inter-related technical and location changes taking place in the car industry are so important that they will result in the emergence of new socio-economic location theory. This proposition can only be tested on the basis of further studies of other industries similarly affected.

The third section points the way towards the research needed to provide a basis for future spatial policy. In demonstrating the possibilities for establishing more decentralized corporate structures in which branch plants possess greater technological autonomy, Charles (chapter 9) makes an important connection between the study of change in organisational, technical and regional systems. Hayter's contribution (chapter 11) is important both in a theoretical and a policy sense. Thus he is at pains to stress that 'mature' industries are not static in a technical sense, although process innovation replaces product innovation as the principal source of change. The charisma of 'high technology' should not lead to the abandonment of 'mature' industries as objects of study. That same charisma should also not be allowed to result in a similar neglect by public agencies. Indeed, Hayter makes a convincing case for a more positive attitude towards 'mature' industries on the part of both the academic community and government. By contrast, the two final chapters stress some of the risks involved in adopting an overoptimistic view of the potential of high technology. They show how the role of the public sector, whether it is heavily involved financially in one industry or in sponsoring or supporting generic initiatives, needs to be re-examined in the light of recent research into high-technology industry. With so much public money being channelled into spatial policies to stimulate or support such industry, this is an important area of research which would help maximize success and minimize waste.

REFERENCES

Freeman, C. and Soete, L. L. G. (1986) *Technical Change and Full Employment*, Blackwell, Oxford.
Malecki, E. J. (1983) Technology and regional development: a survey, *International Regional Science Review* 8, 89–125.

The Contributors

J. BRADBURY John Bradbury is an associate professor of geography at McGill University in Montreal. He has conducted research on industrial geography and regional development in Canada. He has published articles on the Canadian mining industry, Canadian steel industry, resource industrial growth, mining settlements and single enterprise communities.

K. CHAPMAN Keith Chapman is senior lecturer in geography at the University of Aberdeen. His research interests are mainly concerned with spatial aspects of the activities of large companies, especially in the oil and chemical industries.

D. CHARLES David Charles is a research associate at the Centre for Urban and Regional Development Studies, University of Newcastle upon Tyne, with special interests in the electronics industry, research policy and regional development.

C. CHRISTY Craig Christy graduated from the University of Waterloo in 1977 with a Bachelor in Environmental Studies and in 1986 with an MA in Geography from the University of Alberta. Previously a consultant planner, he is now a consultant specializing in high technology industry in Edmonton.

C. FREEMAN Chris Freeman graduated at the London School of Economics. His career has incorporated experience in the army, commerce, education and research. Director of the Science Policy Research Unit, University of Sussex 1966–1981, he is now Emeritus Professor at Sussex working on technical change and economic theory. Main publications are the economics of innovations, technology and employment, and technology policy.

R. T. HARRISON   Richard Harrison is a lecturer in the Department of Business Studies, The Queen's University of Belfast. He has extensive research and publication interests in regional development analysis (including small firms, policy analysis and technical change) and has just completed a Ph.D. thesis on the development and impact of the UK shipbuilding industry.

R. HAYTER   Roger Hayter is an associate professor and chairman of the Department of Geography at Simon Fraser University. He has degrees from the universities of Newcastle (BA), Alberta (MA) and Washington (PhD). His research interests are concerned with the geography of enterprise, especially with respect to the internationalization of firms and the Canadian forest product industries.

J. HOLMES   John Holmes received his doctorate from the Ohio State University in 1974 and is presently associate professor of geography at Queen's University, Kingston, Ontario. His recently published work has focused on various aspects of the political economy of contemporary economic restructuring.

G. HUMPHRYS   Graham Humphrys is senior lecturer in geography and Dean of the Faculty of Economic and Social Studies at the University College of Swansea. His research interests are in industrial development and in northern Canada. His most recent book contributions have dealt with power and the manufacturing industries in Britain, the geography of public sector industries and the industrial development of South Wales.

S. HUSAIN   Sohail Husain has been lecturer in geography at Southampton University since 1976, where his research has been focused on European economic development. He has published articles on structural change in industry and agriculture and, with David Pinder, on oil refining. He is also interested in the environmental impact of economic activities.

R. G. IRONSIDE   Geoff Ironside is a professor of geography at the University of Alberta. He has published extensively on problems of regional and rural development and acted as consultant to federal and provincial governments in Canada.

R. T. McARTHUR   Richard McArthur is a lecturer in geography at Cheltenham College of St. Paul and St. Mary, England. Prior to that he worked as a Research Assistant on an Anglo-French collaborative project funded by the ESRC and CNRS on 'local industrial systems'.

D. PINDER   David Pinder lectures in geography at Southampton University. Since the early 1970s his research and writing have concentrated on economic development in Western Europe. In 1978–9 he was Visiting Lecturer in the Economic Geography Institute of Erasmus University, Rotterdam, and from 1982 to 1986 he was Honorary Treasurer of the Institute of British Geographers.

J. SIMPSON   Jamie Simpson completed his BA and MA at the University of Manitoba and is currently enrolled in the doctoral programme of the LSE. He has jointly authored several articles and books on the aerospace industry and is now engaged in analysis of the regional effects of defence procurement.

D. TODD   Educated at the Universities of Leeds, Queen's (Ontario) and London (LSE), Daniel Todd is a professor in the geography department of the University of Manitoba. His interests lie in the relationships between industrial policy and regional development. He has recently published books and articles on shipbuilding and aerospace.

P. A. WOOD   Peter Wood is senior lecturer in geography at University College London. After earlier work on industrial location and the West Midlands, he has more recently published studies on economic change and planning in and around London, including Docklands, and on the service sector on both sides of the Atlantic.

Index